# The Moon: Co-Author of Life's Script

Sandeep Chavan

Published by SANDEEP CHAVAN, 2025.

THE MOON: CO-AUTHOR OF LIFE'S SCRIPT

**First edition. July 24, 2025.**



ISBN: 979-8231192823

Written by Sandeep Chavan.

# Table of Contents

To the Moon —Silent anchor of our tides,Keeper of rhythms too subtle to measure,You never asked to be worshipped.You only asked to be **seen**.

This book is for you —Not as a mystery to solve,But as a message we finally remembered.

Stillness was always your truth.We just had to stop moving to hear it.

*"The Moon never moved. We just kept circling our own illusions —until her stillness became louder than our science."*

— Sandeep Chavan.

# Disclaimer

This book presents a non-mainstream reinterpretation of celestial phenomena, planetary behavior, and cosmological principles. It is rooted in the framework of **Universal Energy Dynamics (UED)** — an independent, field-based model developed through personal research, observation, and philosophical inquiry.

The ideas expressed herein:

- Depart from classical physics, astronomy, and space science,
- Do not represent the official views of any academic institution, space agency, or scientific organization,
- And are not intended to replace empirically established methods in scientific practice — but to challenge their assumptions and inspire alternative exploration.

While the book may reference historical figures, established theories, and scientific terminology, it uses them as launch points for reframing, not as targets of criticism or dismissal. Respect is given to the thinkers of the past; this work simply asks what they might have missed — or what we might have misinterpreted.

This work is a philosophical and scientific proposition, not dogma. It is not written to convince everyone. It is written for those who are ready to reconsider everything.

Readers are encouraged to:

- Verify observational claims with available data.

- Engage with these ideas critically, openly, and independently.
- Experience the sky themselves — not just through equations, but through alignment.

If this book unsettles your understanding — good. That is not a side effect. It is the intended consequence of thinking.

# Opening Note

*The Moon Did Not Orbit. It Waited.*

We watched the Moon for centuries — charting its phases, mapping its craters, building calendars from its rhythm. It became our first clock, our first mirror, and our oldest companion in the sky.

But as science advanced, we moved from reverence to reason. We called the Moon a satellite. We calculated its orbit. We assigned it spin and force, declared it tidally locked, and wrapped it in the machinery of gravity.

We forgot how to simply look.

For in truth, the Moon never moved. It never rotated, never traveled a track, never obeyed the math we imposed upon it. It simply remained — holding its face toward us with unwavering stillness.

Not out of coincidence. Out of alignment.

This book is not just about the Moon. It is about everything the Moon reveals — when we stop assuming motion and start understanding resonance.

It is about a universe that does not spin, but pulses. A planet that does not orbit, but stabilizes. A reality not of forces and speeds — but of ripples, harmonies, and silent consequences.

The Moon did not orbit to please our physics. It aligned to sustain our life.

And now, as we learn to observe again — through stillness rather than equations — we begin to see the Moon for what it truly is:

Not a rock in space. But the co-author of Earth's story. The ripple guardian of our rhythm. The quiet proof that motion was never needed — only meaning.

Welcome to a universe where stillness speaks louder than spin. And the Moon... finally gets the credit it never asked for.

# How to Read This Book

This is not a book about the Moon. It is a book through the Moon.

It will not walk you through textbook astronomy or scientific consensus. Instead, it invites you to pause — and perceive the sky not as an object of observation, but as a mirror of alignment.

**Read it like a conversation.**

Each chapter builds a shift in perspective. Don't rush. Reflect. Let your own experiences of the Moon — your memories of still nights, changing tides, strange dreams — join the dialogue.

**Read beyond the words.**

This book is written in a rhythm. The structure, the pauses, the repetitions — all are intentional. They echo the very pulse it tries to describe. You are not just absorbing ideas; you are tuning into a pattern.

**Question everything you "know."**

You'll encounter ideas that contradict what you were taught — about rotation, orbit, gravity, even time. Don't seek immediate agreement or rejection. Suspend judgment. Let these views sit beside the ones you hold. Watch what shifts.

**Come back to it.**

Like the Moon, this book reveals differently depending on when and how you return. What doesn't make sense today may ripple into clarity tomorrow — especially as your perception of motion, stillness, and structure begins to re-align.

**Use the Moon as your guide.**

Don't just read indoors. Step outside. Look at the Moon. Observe its face. Its silence. Its rhythm. Let it show you what this book cannot.

This book is not about proving a theory. It is about remembering a resonance.

The Moon never demanded your belief It simply held its place — until you were ready to see it.

Begin anywhere. But read with stillness. Because that's where everything begins.

# Preface

*The Mirror in the Sky*

We live in a world addicted to motion — a culture that celebrates speed, movement, and acceleration as evidence of progress. Every scientific theory, from Newton's gravity to Einstein's relativity, has been constructed upon the foundation of movement. Orbits, spins, expansions, collisions — these are the metaphors we've used to describe the cosmos, and in turn, ourselves.

But what if we were wrong?

What if the universe was never truly in motion — only our perception was?

What if all that we assumed to be rotating, revolving, expanding, and collapsing... was simply resonating in stillness?

The Moon, our most loyal celestial companion, has been silently holding that truth for us all along.

It has shown the same face for as long as we've looked up — unmoved, unspun, unwavering. While the Sun blinds us with brilliance and the stars overwhelm us with scale, the Moon simply reflects. It reflects not just sunlight, but something far more subtle — our assumptions about reality.

We built entire models of physics by watching the Moon. Newton used it to prove gravity. Einstein folded it into the curvature of spacetime. Astronomers used it to predict eclipses and calculate tides. Poets used it to inspire emotion. Prophets marked holy days by its crescent.

But few ever paused to ask the most obvious question: Why does the Moon never move — and yet, everything moves because of it?

This book is that pause.

It is a return to first observation — not through inherited models, but through honest perception.

It is a challenge to motion-based cosmology and an invitation to consider a universe governed not by mechanics, but by structured equilibrium. Not by push and pull, but by ripple consequence. Not by gravity, but by resonance.

The Moon did not orbit for Newton. It did not spin for science. It stayed — in alignment, in posture, in presence. And in doing so, it quietly shattered the illusion of motion we built around it.

In its stillness, it extended an invitation:

"Look again. Not with equations, but with awareness. Not through the speed of things, but through the harmony of their presence."

This book accepts that invitation.

# Part I – The Great Assumption

# Chapter 1: The Silent Companion

*Rediscovering the Moon's historical, cultural, spiritual, and functional significance — not as a rock in orbit, but as a resonant presence guiding human consciousness, timekeeping, and meaning-making for millennia.*

We did not build monuments to the Sun. We survived under it, labored beneath it, feared its droughts and welcomed its warmth. But when we gazed upward at night and saw the Moon—glowing softly, changing gently, yet never leaving us—it was then we began to understand time, rhythm, and reflection.

Before we could write, we could count the days by the Moon. Before we could navigate with instruments, we followed its waxing and waning light through forests, oceans, migrations, and myths. The Moon was our first teacher, our first clock, our first mirror in the sky.

It asked for nothing. It simply stayed.

While the stars wheeled across the heavens and the Sun dominated the day, the Moon whispered a quieter truth—one not of domination or distance, but of alignment and presence. It shaped the earliest calendars, marked sacred days, guided the tides, and reflected back to humanity its own cycles of longing, change, and renewal.

And yet, in our obsession with equations and orbits, we reduced this silent guardian to a "natural satellite" — a cold, obedient rock circling our planet by the laws of Newton and chance.

But the ancients knew better.

They did not measure the Moon. They listened to it.

## 1.1 Introduction – A Companion, Not a Machine

We always looked up and saw the Moon — and it always looked back.

Not with fire, not with fury, but with calm. With rhythm. With presence. In every culture, in every era, across every continent and civilization, the Moon was never a stranger. It was the first familiar — the constant companion in the sky that followed us across mountains and deserts, above cities and rivers, in solitude and celebration.

It didn't rise with noise or light. It simply appeared — as if to remind us that something above us was still watching, even when the Sun had gone. Unlike the stars that scattered endlessly, or the Sun that came and went with harsh finality, the Moon lingered, moved gently, changed slowly, and returned faithfully.

It was never feared. It was felt.

Long before we built observatories or launched satellites, long before Newton imagined its orbit or Einstein curved its path, the Moon had already entered our stories, our prayers, our rhythms. It was marked on bones, etched into stone, sung in lullabies, and studied by priests and poets alike. We used it not just to mark time, but to make sense of ourselves.

This was not a rock. This was a guidepost.

The Moon gave us more than light at night. It gave us structure in a world full of uncertainty. It marked beginnings and endings. It told us when to sow and when to rest. When to fast and when to feast. It did not calculate — it communicated.

But then came the age of reduction. We turned this companion into a machine. We called it a "satellite." We mapped it, measured it, and stuffed it into our formulas. We declared it to be rotating, orbiting, and tidally locked — all based on assumptions stitched into the fabric of Newtonian mechanics.

We stopped listening. We started modeling.

Yet through it all, the Moon never argued. It never moved faster to prove us wrong or glowed brighter to reclaim attention. It simply continued — holding the same face, in the same posture, with the same unwavering ripple of stillness.

That's the premise of this chapter, and of this book:

The Moon was never a passive object. It was — and still is — a resonant presence, silently guiding humanity through the illusions of motion, back to the deeper truth of alignment, rhythm, and consequence.

Before we calculated its orbit, the Moon had already calibrated our lives. Now, it's time to return to that stillness — and listen again.

## 1.2 The Moon in Ancient Civilizations

To truly understand how humanity once perceived the Moon, we must look beyond physics — into ritual, rhythm, and reverence. For thousands of years, across every corner of the Earth, the Moon was not treated as an object of study — it was regarded as a divine participant in life itself.

It wasn't analyzed. It was consulted. It wasn't merely observed — it was honored.

**Egypt: Thoth and the Calendar of the Heavens**

In ancient Egyptian cosmology, the Moon was inseparable from Thoth, the god of wisdom, time, writing, and equilibrium. Thoth was said to have measured the heavens, regulated the stars, and brought balance between chaos and order — always with the Moon as his divine instrument.

The Egyptian lunar calendar was one of the earliest formal attempts to align time with celestial rhythm, especially with respect to agriculture, festivals, and the afterlife.

The Moon, in this context, was not a secondary object.

It was the celestial ruler of cycles, assigned to mark sacred time — when the veil between the earthly and the divine was thinnest.

**Babylon: Eclipses, Omens, and Lunar Sovereignty**

The Babylonians elevated the Moon to the rank of a god: Sin (or Nanna) — the "lord of wisdom" and father of the Sun god Shamash. They meticulously tracked lunar eclipses, not as astronomical phenomena but as cosmic warnings and divine omens that could shift the fate of kings and empires.

Farmers looked to the Moon to decide when to plant or harvest. Priests used its phases for divination and ritual precision. Babylonian tablets from over 2,000 years ago contain exact lunar calculations, yet not once do they treat the Moon as mechanical. It was alive in meaning — a celestial will expressing itself through shadow and shimmer.

**Maya and Aztec: Calendars of Light and Shadow**

Few civilizations revered the Moon with as much mathematical sophistication and spiritual commitment as the Maya and Aztecs. Their calendars weren't simply tools of time — they were sacred maps of cosmic order.

The Maya lunar calendar tracked cycles with astonishing accuracy, integrating them into agricultural rituals, royal ceremonies, and spiritual initiations. Lunar eclipses were carefully recorded and spiritually interpreted. The Moon determined when the community would gather, fast, dance, or mourn.

To the Aztecs, the Moon had its own creation myth, and its phase changes were symbolic — not just of time passing, but of divine drama unfolding above the Earth. Waxing was growth. Waning was descent. The Moon didn't just tell time — it *shaped it.*

**China: Unity Under the Full Moon**

In Chinese cosmology and tradition, the Moon carries the weight of unity, completeness, and reunion. The Mid-Autumn Festival, one of the most beloved festivals in Chinese culture, is centered around the full Moon — symbolizing family togetherness, harmony, and gratitude.

Ancient Chinese astronomy tracked lunar phases with great care, embedding them in both imperial rituals and daily life. The Moon was considered a living balance to the Yang energy of the Sun — a feminine Yin force that cooled, soothed, and stabilized. Poets wrote of longing under moonlight, lovers exchanged promises by it, and emperors offered prayers in sync with its cycles.

**Greece and Rome: The Divine Feminine in the Sky**

In Greco-Roman mythology, the Moon was personified as a goddess: Selene to the Greeks and Luna to the Romans. She was portrayed riding a silver chariot across the night sky, her light illuminating more than just darkness — it awakened dreams, desires, and destinies.

Selene was not merely the Moon — she *was* the Moon. Divine, feminine, ethereal. She governed the tides of both oceans and emotions. Luna's influence extended to medicine, philosophy, and statecraft. In Roman temples, Moon phases were synchronized with sacrifices, laws, and public rituals.

**A Common Thread Across Time**

Across cultures and continents, we find a repeating pattern:

- The Moon was not just light — it was law.
- Not just presence — but power.
- Not just moving — but meaningful.

And it was never called a satellite. Never considered a dead stone in the sky. It was always alive — with intention, with rhythm, with voice.

The ancients didn't need telescopes or gravity equations to know that the Moon mattered. They felt its ripple long before we thought to measure it. It ordered their world and stabilized their spirit.

What they saw through reverence, we are only now beginning to rediscover through science — That the Moon was never moving.

It was simply resonating. And everything on Earth — from crops to cultures to consciousness — followed its silent script.

## 1.3 The Moon in Religious Symbolism

If civilizations honored the Moon as a celestial guide, religions entrusted it with divine authority. The Moon was never merely a calendar tool or aesthetic muse — it was woven into the sacred, entrusted with regulating human spiritual alignment with the cosmos.

In every major tradition, the Moon is the keeper of sacred time — not because of its shape or orbit, but because of its resonance with human experience. It marks beginnings and endings, fasts and feasts, reflection and renewal. The Moon's cycle was seen not as a neutral observation but as divine instruction, a heavenly clock ticking with cosmic intention.

Let's explore how the Moon became spiritually indispensable across religious traditions.

### Islam: The Crescent that Opens Sacred Time

In Islam, the Moon is the cosmic signal for divine timing. The Hilal — the first visible crescent — is not merely observed; it is awaited, with spiritual anticipation. This delicate sliver of light marks the beginning of each lunar month, including the sacred month of Ramadan and the joyous celebrations of Eid al-Fitr and Eid al-Adha.

The Prophet Muhammad (peace be upon him) taught:

*"Fast when you see it, and break your fast when you see it."* — reinforcing that spiritual cycles begin not by human will, but by cosmic alignment.

In UED terms, Hilal isn't just a visible curve of light — it's the first threshold of ripple resonance between Earth, Moon, and human perception. Its appearance isn't motion; it's permission. Permission to enter a new sacred phase, with body, mind, and spirit recalibrated to celestial rhythm.

The Moon, thus, governs not just time in Islam — it governs divine alignment.

**Hinduism: Chandra, Nakshatras, and the Inner Cosmos**

In Hinduism, the Moon is not a background object. It is a deity — Chandra — radiant, gentle, and essential in the vast network of cosmic order. He is often depicted riding a chariot across the sky, pulled by ten white horses, shining with cool lunar grace that balances the solar fire of Surya.

The Moon is central to Jyotish Shastra (Vedic astrology), where it travels through 27 Nakshatras — lunar mansions — each governing emotional, psychological, and energetic patterns in life. While Western astrology prioritizes the Sun sign, Hindu astrology centers the Moon sign, acknowledging its direct impact on human mind (manas), emotions, and subconscious patterns.

Many Hindu rituals — from Ekadashi fasting to Purnima (Full Moon) and Amavasya (New Moon) observances — are precisely synchronized with lunar phases, not for tradition's sake, but to optimize inner clarity and energetic purity.

In UED logic, these practices are not mystical folklore — they are intuitive field-tuning rituals, synchronized with the Earth-Moon-Sun ripple alignment.

**Judaism: The Moon as the Keeper of Sacred Memory**

The Hebrew calendar is entirely lunar, beginning each month with the new Moon (*Rosh Chodesh*) and marking sacred festivals by its phases. Passover — the foundational narrative of Jewish liberation — is observed during the full Moon of Nissan, symbolizing clarity, fullness, and divine intervention.

The Moon's waxing and waning served as spiritual metaphor for the rise and fall of Israel, the exile and return, the loss and restoration of divine presence. The Moon, in this context, was not just a timekeeper — it was a symbol of hope, a visible cycle of divine concealment and revelation.

Rabbis often spoke of the Moon as God's faithful servant, the lesser light that reflected glory and humility — the hidden greatness of spiritual life.

**Christianity: Resurrection by the Moonlight**

While Christianity later adopted the solar Gregorian calendar, it still retains deep lunar undercurrents. The most sacred Christian celebration — Easter — is calculated using lunar alignment: it falls on the first Sunday after the first full Moon following the March equinox.

This is not a coincidence — it is rooted in the early Church's Jewish origins and cosmological awareness. The full Moon was seen as the divine spotlight, illuminating resurrection, renewal, and spiritual rebirth.

Monastic communities in medieval Europe, too, followed lunar cycles for prayer, fasting, and observance. Their connection to lunar time kept them attuned to silence, reflection, and divine rhythm — long before mechanical clocks standardized artificial time.

**Buddhism: Enlightenment Under the Full Moon**

In Buddhism, the Moon is a symbol of awakening — serene, detached, and illuminating the mind.

The full Moon is deeply sacred in Theravāda tradition, where Uposatha days (Buddhist sabbaths) align with it. On these days, monks and lay followers intensify their meditation, moral discipline, and reflection.

In Sri Lanka, the Poya day — always observed on the full Moon — marks key events in the Buddha's life: his birth, enlightenment, first sermon, and passing. The Moon here is not merely a celestial light — it is a mirror for the mind. Still, bright, and detached from emotion — the Moon symbolizes the state of Nirvana.

In UED terms, the full Moon represents the maximum phase of alignment — both in sky and in consciousness.

A moment when ripple resonance is purest, enabling inner clarity.

**The Sacred Resonance: A Universal Pattern**

Whether it was called Hilal, Chandra, Luna, Selene, or Sin, the Moon was not classified — it was revered. Every tradition recognized the same truth through different names:

- The Moon was not a clock. It was conscious time.
- It did not move — it moved us.
- It did not demand — it reflected.
- And always, it waited — for our spirit to align with its rhythm.

In UED perspective, this universal reverence wasn't superstition. It was early humanity's innate resonance with cosmic stillness — an intuitive grasp of ripple logic long before it was written in equations.

The Moon, across every sacred text and holy path, was never reduced to data. It remained what it still is —

A divine alignment. A cosmic compass. A silent messenger of deeper realities.

## 1.4 The Moon as Humanity's First Timekeeper

Long before pendulums swung, sands slipped, or gears turned, time was not measured — it was experienced. And the Moon was the first hand on that cosmic clock.

In a world without digital calendars or mechanical watches, early humans lifted their eyes to the sky, and the Moon offered them something no other celestial body could: visible, rhythmic change.

Predictable. Gentle. Alive.

In the absence of artificial light and precision instruments, the Moon's ripple rhythm became the heartbeat of human activity — spiritual, social, agricultural, and biological.

**The First Calendars Were Carved in Bone, Not Bronze**

Archaeologists have uncovered evidence that humans were tracking the Moon at least 25,000 years ago. The famous Ishango bone from Central Africa — dated to around 20,000 BCE — contains tally marks widely interpreted as a lunar calendar, possibly used by women to track menstrual or fertility cycles.

Similar etchings found in paleolithic caves in France, Germany, and Turkey show groups of dots or lines arranged in consistent patterns — correlating not to solar movement, but to the 29.5-day lunar cycle.

These weren't decorations. They were records of resonance — the earliest observable alignment between nature and experience. In UED terms, they were humanity's first efforts to mark ripple consequence across time.

**The Moon and the Feminine Cycle: Cosmic Synchrony**

One of the most profound biological alignments is that of the human menstrual cycle with the lunar month — both averaging approximately 29.5 days. This synchrony is not coincidental. It's resonant.

Across ancient cultures, this correlation was honored, not hidden. Women were seen as living calendars, mirroring the waxing and waning of the Moon. Fertility, birth rituals, and sacred feminine rites were tied to lunar phases, particularly the new moon (conception/renewal) and full moon (ripeness/enlightenment).

In ripple dynamics, this isn't symbolic. It is the biological entrainment of human tissue to the ambient lunar field —an echo of the Moon's steady presence on Earth's internal rhythm.

**Lunar Light: The Original Guide for Survival**

For early humans, lunar phases dictated when to plant, when to harvest, when to move, and when to stay. The Moon determined more than mood — it regulated migration routes, trade paths, animal behavior, and water patterns.

- **Farming**: Farmers sowed crops based on the waxing moon (growth energy) and harvested under the waning moon (release).
- **Migration**: Hunters and nomads timed their travel based on the full moon's light, giving them safer passage during dark nights.
- **Trade**: In coastal and river-based cultures, tidal awareness governed by lunar pull directly influenced boat movement and marketplace timing.

The Moon's cycle became a social contract with nature — an agreement to move in harmony with cosmic rhythm, not against it.

**The Full Moon: Earth's First Nightlight**

Before firelight, before candles, before the torch or the oil lamp, the full moon was Earth's only reliable nocturnal illumination. It bathed the landscape in silver clarity, revealing paths, predators, and possibilities.

Communities often gathered under full moons to celebrate, perform rituals, conduct storytelling, and make communal decisions. In many early traditions, marriages, births, and important announcements were aligned with lunar brightness, enhancing both symbolism and safety.

The full moon was not just light —It was a temporal signal, a ripple peak, a field amplifier for shared experience.

**UED Perspective: Ripple Rhythm Before Chronology**

Before mechanical clocks imposed artificial time structures, the Moon offered a natural resonance — a cosmic beat that informed human biology and behavior. Time wasn't external — it was embodied.

- The phases were not divisions, they were consequences.
- The darkness of the new moon wasn't absence — it was

overalignment.
- The crescent was not emerging — it was crossing the threshold of perception (ATT).

The Moon taught us time without ever teaching it. It offered order without enforcement, structure without rigidity, and stillness that moved all things.

**Final Thought:**

"We did not invent time. We aligned with it. And the Moon was the first to show us how."

Long before we counted seconds, we followed shadows. Before we wore watches, we followed whispers in the sky. And the Moon — silent, patient, unwavering — became our first invitation to live with the universe, not just inside it.

## 1.5 From Reverence to Reduction

For most of human history, the Moon was not just a light in the sky — it was a presence, a companion, a mirror of meaning. It told stories, guided decisions, marked sacred days, and reflected the rhythms of the Earth and the human body alike. It was not questioned — it was trusted.

But then, something changed.

As the Age of Reason dawned, so too did a new way of seeing. Telescopes replaced naked eyes. Calculations replaced ceremonies. Observation became measurement. Reverence became curiosity. And the Moon — once a divine partner in humanity's spiritual and temporal life — was downgraded to a machine-driven object in an impersonal system of motion.

**The Rise of Mechanistic Cosmology**

When thinkers like Copernicus, Kepler, Galileo, and Newton began to explore the heavens, they did so with sharp intellect and revolutionary courage. They dethroned geocentrism and introduced an elegant framework where bodies moved in predictable paths, governed by universal forces.

In this shift, the Moon became the first celestial body to be reduced to mechanics:

- It was no longer a goddess or guide, but a satellite.
- Its phases were not signs, but angles of reflection.
- Its rhythm was not sacred, but calculated.
- Its alignment was not meaningful — just coincidental.

And in doing so, the Moon's relational significance was erased — replaced with a cold, orbital chart.

The sacred light that once illuminated our collective memory was converted into a point mass moving under gravity.

**From Ripple to Formula: The Loss of Dimensional Depth**

Newton's law of gravitation explained the Moon's apparent behavior using mathematics. The Moon, he argued, was constantly "falling" toward Earth but missing it due to forward momentum — thus entering a perpetual orbit.

This idea was brilliant — but incomplete.

It worked on paper, but failed to account for perception. For in truth, the Moon never appears to rotate, never reveals its dark side, and shows no visual evidence of being dragged in a mechanical dance.

Still, the equations took over. Science, enchanted by predictability, adopted the gravitational model wholesale. The Moon was assigned an orbit, a rotation, and a lock — not because we observed it directly, but because the math required it.

In this model, the Moon was not listening to us, it was obeying the laws of force.

And thus began the era of reduction.

**The Fall from Mirror to Machine**

The reverent Moon — the one worshipped by sages and tracked by tribes, the one that sang in stories and whispered through rituals — was reclassified:

- From divine feminine to rocky remnant
- From resonant guide to gravitational hostage
- From harmonic mirror to mechanical sphere

Telescopes captured craters. Satellites measured surface temperatures. Space probes landed on dust. And in all this, the Moon's silence was mistaken for compliance.

What was once interpreted as stillness became tidal locking. What was once seen as presence became probability. What was once an echo of universal balance became an obedient equation.

We stopped asking what the Moon meant. We only asked where it was going, how fast, and why.

**UED Reflection: From Faith to Friction**

From the UED perspective, this shift from reverence to reduction marked the loss of intuitive resonance. Humanity moved from being participants in ripple harmony to becoming external modelers of objects.

We no longer saw ourselves as aligned with the Moon — we saw ourselves as observers of it. Yet, the Moon never changed. It stayed still. It continued to align. It refused to perform motion, even as we insisted on describing it in those terms.

This chapter in history was necessary — but incomplete. It helped us discover many things — but blinded us to what was always right in front of us.

**Final Thought:**

*"The Moon did not descend — our perception did. We stopped dancing with it, and began dissecting it."*

But now, the invitation returns:

To re-honor the Moon not with myth, but with meaning.

To reframe it not as a machine in orbit, but as a ripple-stabilizer in resonance.

To move from measurement back to alignment.

## 1.6 UED Reflection: Why the Moon Stayed Constant

If science has tried to explain the Moon through orbit, then UED reclaims it through alignment. Because when we strip away inherited models and observe with clarity, one undeniable truth remains:

The Moon doesn't spin. It doesn't change its face. It doesn't chase or flee. It stays.

And in that stillness, it does more than reflect light — it stabilizes life.

**Stillness Is Not Stagnation — It's Harmony**

In most scientific thinking, stillness is a default — an absence of force, a lack of movement. But in UED, stillness is not the absence of action; it is the resolution of tension, the achievement of equilibrium, the silent harmony of all conflicting ripple forces balanced into form.

The Moon's still face is not a flaw in our observation. It is the first proof of this deeper ripple architecture.

It is not frozen — it is resolved.

It's not obeying gravity — it's anchoring resonance.

While the Earth pulses with weather, seasons, tectonics, and turbulence, the Moon remains the unmoving observer in our sky — the one stable ripple mirror in a field of chaos.

**Ripple Alignment, Not Orbital Motion**

From the UED lens, the Moon isn't orbiting Earth — it is resonating with it. What we perceive as movement is merely a consequence of shifting ripple contrasts in the Structured Vacuum

Energy (SVE) field. The Moon doesn't "go around" — it locks in place through harmonic alignment.

Its face remains constant not because it is spinning just right, but because its field posture is ripple-stable. It is the cosmic equivalent of a tuning fork held perfectly to tone — not vibrating wildly, but sustaining the note of balance.

In mechanical models, this looks like tidal locking.

In UED, this is consequence without displacement — the Moon as a standing ripple node, aligned precisely between Earth and Sun in a way that requires no motion to maintain its role.

The Moon does not move to stay aligned — It stays aligned to avoid the need to move.

**The Moon as Earth's Ripple Anchor**

The Earth, too, is not spinning wildly through space in UED. It exists in its own ripple equilibrium, constantly exposed to ripple pressure from the Sun. This pressure — in wave models — would cause turbulence, instability, even collapse of climatic and biological systems.

Enter the Moon — not as a gravitational satellite, but as a ripple buffer, an interference stabilizer.

It absorbs misaligned solar ripple tensions. It deflects overflow ripple intensities. It holds Earth's field in coherence like a cosmic counterbalance — a silent stabilizer in a turbulent ripple sea.

Without the Moon, Earth's field would be more volatile, more chaotic, less capable of sustaining harmonic life.

**The Moon as the First Stabilizer of Consciousness**

Before civilization, before agriculture, before even language — the Moon was already teaching us something vital: consistency.

Its steady phases, its return, its rhythm — they became the first outer mirror for inner perception. It offered a sense of time, but more deeply, a sense of trust.

- Trust that something in the cosmos does not change.
- Trust that life moves in patterns.
- Trust that stillness can coexist with transformation.

In UED terms, the Moon was humanity's first long-term ripple alignment — a consciousness anchor that offered psychological grounding long before we could even name it.

It didn't need to speak. Its stillness was the message.

While the stars scattered, the Sun blinded, and the Earth shifted — the Moon remained our constant ripple companion.

**Final Realization:**

*"Stillness is not the opposite of change — it is the space in which change becomes meaningful."*

The Moon never had to rotate. It never had to move. It simply aligned — and by doing so, became the stabilizer of Earth's balance, and the seed of human reflection.

In stillness, it gave us time. In rhythm, it gave us order. In presence, it gave us meaning.

## 1.7 Closing Thought: The Eye in the Sky

There was a time when humanity didn't ask how the Moon moved — because it didn't need to.

It was enough that it appeared. That it returned. That it glowed when the world grew dark. The Moon wasn't something to conquer or decode — it was something to follow.

Before astronomy charted it, before mathematics tried to own it, the Moon was something we experienced — not intellectually, but intimately. It spoke in silence. It didn't demand explanation; it offered rhythm. Its phases told us when to gather, when to let go, when to rest, and when to begin again.

We didn't measure the Moon. We listened to it.

Long before orbit maps and gravitational equations, we simply observed its rise and fall. Its crescent meant potential. Its fullness meant clarity. Its disappearance meant pause. These meanings weren't imposed — they were felt, known, remembered. Not with data, but with presence.

We didn't track its orbit. We moved with its rhythm.

And when we looked up, night after night, and saw the same face gazing back — we recognized something rare in nature: constancy. A cosmic calm. A reliable reflection in an otherwise shifting world.

The Moon was never just a stone circling our sky. It was, and still is, a mirror — not of sunlight alone, but of our human longing for order, rhythm, and stillness.

It was not a rock. It was a reflection.

In its still face, we did not find just light. We found meaning. A quiet reassurance that in the midst of all uncertainty, something remained unchanged. A ripple of permanence in a sea of chaos.

And perhaps, most importantly:

The Moon reminded us that not all truths must move to be real. Some simply need to remain.

# Chapter 2: What We See, What We Were Told

*The tension between observable lunar stillness and the theoretical machinery imposed on it by classical science. This chapter exposes the visual paradox and frames the historical rise of orbital dogma.*

We have been looking at the Moon for thousands of years — and it has never blinked.

Its craters, its shadows, its scars — all stare back at us with unshakable consistency. Night after night, month after month, era after era — it shows us the same face. Not a slightly different angle. Not a rotating canvas. The *same* face.

And yet, modern science tells us otherwise.

We are told the Moon is spinning. Rotating on its axis. Orbiting at over 3,600 kilometers per hour. Perfectly synchronized, we're assured, so that it always shows us the same side — a gravitational coincidence, locked in time.

A neat theory. A mathematical marvel. And a complete contradiction of what we actually see.

This chapter is about that contradiction.

It is about what we've observed for millennia — and what we've been told to believe in spite of it. It is about the moment our ancestors' intuition was overruled by orbital equations, and how we traded stillness for spin, alignment for mechanics, truth for theory.

It is also about the courage to look again — not through inherited explanations, but through the eyes that never stopped seeing.

The Moon didn't stop moving. It never moved to begin with. We just stopped trusting what it was trying to tell us.

## 2.1 Introduction – The Moon That Doesn't Budge

It's one of the most consistent experiences in all of human history: we look up — and the same face of the Moon greets us.

The same craters. The same scars. The same silent gaze.

For millennia, civilizations across continents, with no knowledge of one another, all saw the same Moon. Pharaohs, monks, warriors, farmers, poets — all looked up at night and shared a common companion in the sky. It didn't rotate in front of their eyes. It didn't shift its posture. It didn't reveal a new angle the next day or year or century.

It stayed still. Not figuratively — visibly.

And yet, in the age of modern science, we are told with absolute certainty: The Moon rotates. It spins once every 27.3 days. Perfectly matched to its orbit around Earth. So perfectly, in fact, that we never notice it turning.

We're told this is called tidal locking — a gravitational phenomenon where rotational and orbital periods become synchronized. And just like that, what looks like stillness becomes a story of motion so perfect you can't detect it.

But this raises a fundamental question — one too often ignored in textbooks and lecture halls:

Why, after thousands of years of observation, do we still insist that something is moving when we've never seen it move?

This chapter is about that paradox. The tension between direct experience and inherited theory. Between what the Moon shows us and what we are told to believe about it.

It asks not just whether the Moon is rotating — but why we were so eager to explain its stillness away.

## 2.2 The Visual Truth: What the Naked Eye Tells Us

The Moon has no secrets — it has been openly visible every night of human existence, quietly presenting itself to every generation, civilization, and observer who cared to look. And what it shows us is simple, stunning, and unwavering:

It never changes its face.

Whether you observe it from the deserts of Arabia, the temples of Varanasi, the savannahs of Africa, or the mountains of Peru — the face of the Moon remains the same. The great dark plains known as the mare, the deep Tycho crater, the lighter highlands — all hold their positions as if etched permanently into a cosmic canvas.

**A Constant Portrait, Not a Rotating Globe**

This isn't poetic exaggeration — it's empirical consistency. For thousands of years, no matter what month or season, no matter the political era or technological age, we have seen the same side of the Moon. Not part of it. Not most of it. But exactly the same hemisphere, locked in its position, as if it were a framed portrait in the sky.

- No gradual turning.
- No shifting orientation.
- No progressive unveiling of a "dark side."
- Just the same cratered visage, faithfully aligned with Earth.

In our everyday experience, anything that rotates reveals a new side.

A spinning coin shows its edge, a turning wheel completes a cycle, a rotating globe slowly brings different continents into view.

But the Moon? It does none of these things.

**No Drift, No Change, No Clue of Spin**

If the Moon were truly rotating on its axis, even perfectly synchronized with its orbit, we would expect some observable indicators — however subtle. Over long periods, we should detect a lateral drift, a micro-shift in alignment, or even a slight change in the angle of features visible from Earth.

But we don't.

Even over decades of observation, the Moon holds its posture with eerie precision. Long-exposure photographs. Telescopic tracking. Satellite imaging. Generations of observation by astronomers and amateur skywatchers alike — all confirm the same astonishing fact:

The Moon doesn't rotate in any observable way. It simply remains. Facing us. Unmoved.

**Time-Lapse Confirms the Truth**

Modern time-lapse photography has made this truth undeniable. When we compile thousands of images of the Moon taken night after night, month after month, and stitch them together into motion — there is no motion.

Instead of spinning like a globe, the Moon behaves like a photograph taped to a moving window. The background may shift slightly — due to atmospheric conditions or viewing angle — but the Moon itself? It holds. It stares. It waits.

Even the minor "wobble" known as libration, which science often cites as evidence of subtle movement, turns out to be a consequence of viewing angle and Earth's tilt, not actual rotation. It's a visual illusion — a perceptual shimmer — not proof of axial spin.

**The UED Implication**

From a UED perspective, this constancy is not an anomaly to be patched with tidal locking math. It is a primary clue — a direct observation pointing to a ripple-based resonance, not mechanical rotation.

The Moon's posture isn't "perfectly timed" — it's field-aligned.

Its stability is not a result of force balance — it's the result of stillness in ripple equilibrium.

**Final Reflection:**

"The Moon never turned away. We just turned our theories against what it was showing us."

This is the visual truth. Not mathematical. Not theoretical. Just faithful, consistent, undeniable perception.

And the moment we stop explaining it away is the moment we start listening again.

## 2.3 The Mainstream Explanation: Tidal Locking

Faced with the obvious stillness of the Moon — the same face, night after night — mainstream science doesn't deny the observation. Instead, it reinterprets it through a framework that fits its long-standing mechanical worldview.

The dominant explanation is known as tidal locking — a term that has become scientific gospel, printed in textbooks, taught in classrooms, and accepted by nearly everyone without question.

**The Classical Claim: A Perfect 1:1 Ratio**

According to this theory, the Moon does rotate on its axis, but it does so at exactly the same rate that it orbits the Earth — once every 27.3 days. This synchronized spin and orbit, the theory says, is why we always see the same hemisphere. The rotation just happens to cancel out the orbital motion from our perspective.

The result? The illusion of stillness.

What we see as a Moon locked in place is, by this reasoning, actually a perfectly timed spin — one that just so happens to match its journey around Earth with flawless precision.

A mechanical miracle. A mathematical coincidence. A cosmic tie game.

**The Origin Story: Gravitational Friction as the Sculptor**

The theory further posits that this synchronization wasn't always the case. Billions of years ago, the Moon supposedly spun at a different rate, but over time, Earth's gravitational pull created tidal bulges on the Moon. These bulges caused internal friction, which gradually slowed the Moon's rotation until it matched the pace of its revolution.

Once this 1:1 ratio was reached, the system supposedly settled — and stayed there. Locked. Balanced. Spinning just enough to hide its spin.

Thus, tidal locking became the official resolution to the visual paradox.

**Mathematically Neat, Visually Absent**

The tidal locking explanation is elegant on paper. It creates a simple equation that neatly connects rotation, revolution, and gravitational force. It fits perfectly within Newtonian mechanics and Keplerian orbital models.

But here lies the core problem:

It works in theory — but not in sight.

No one has ever observed the Moon's rotation. No instrument, no time-lapse sequence, no long-term observation has shown the Moon turning in real-time. Not even a whisper of axial movement has emerged across centuries of collective human gaze.

There's no drift. No incremental shift. No slow unfolding of hidden terrain.

Only justifications for why we see none of this.

**The Irony of Perfection Without Evidence**

The most puzzling part of the tidal locking theory is its absolute perfection.

It's not "close enough" — it's exact. The Moon's rotation matches its orbit so precisely that we only ever see 50.1% of its surface due to slight observational wobble (libration). We never get even a fraction more.

This would require extraordinary long-term precision in mass distribution, rotational torque, and orbital energy — all maintained for billions of years, without any measurable decay or drift.

It is, in essence, too perfect.

Too perfect to challenge. Too perfect to test. Too perfect to believe — if one's eyes are still open.

**UED Observation: Coincidence or Consequence?**

From a Universal Energy Dynamics perspective, tidal locking isn't just flawed — it's a conceptual misfire. It tries to force motion into stillness, because the framework of science demands movement to explain stability.

But UED reframes this:

- The Moon does not rotate.
- It does not revolve in the classical sense.
- It maintains ripple alignment in a still universe, held not by gravitational tug-of-war, but by resonant equilibrium in Structured Vacuum Energy (SVE).

Tidal locking, in this view, is not a frictional process — it's an illusion of theory, built to preserve a mechanical model that cannot accept the reality of stillness.

**Final Reflection:**

"Tidal locking explains everything — except what we actually see. Stillness stares us in the face. Tidal math looks the other way."

So we're left with a beautiful theory, perfectly balanced — and entirely invisible.

A story that explains why we never see the Moon move — by insisting that it's moving just right so we never see it.

Science calls that elegance.

Observation calls it evasion.

UED calls it a misreading of resonance.

## 2.4 The Rise of Orbital Dogma

When we trace the roots of modern astronomy and physics, we find that one man sits at the foundation of the structure we still inhabit today: Sir Isaac Newton.

Newton's genius was undeniable. His laws of motion and universal gravitation offered a powerful new lens through which to view the cosmos — one that made the heavens seem calculable, predictable, mechanized. But in building that framework, Newton needed a test case — a real-world example that could demonstrate his mathematical model.

He chose the Moon.

**The Moon as the First Proof of Gravity**

Newton's idea was elegant in its simplicity:

The same force that pulls an apple to the ground must also be acting on the Moon.

He envisioned the Moon as an object constantly falling toward Earth, but missing it due to its sideways momentum — a kind of endless cosmic freefall. In this view, the Moon wasn't suspended by magic or divine harmony — it was being held in orbit by gravity, pulled toward Earth while traveling fast enough to never crash into it.

This became the classic curved trajectory model — the Moon "falling around" Earth in a perfect, silent loop.

It was the first time a celestial motion was described not by divine will, but by mechanical cause. And the Moon, ironically still in our sky, became the symbol of Newton's vision of a universe in motion.

**Why Newtonian Mechanics Needed Motion**

For Newton's model to work — for gravity to be universal — the Moon had to move.

The equations demanded motion. The entire idea of force, acceleration, and inertial balance rested on bodies being in motion.

If the Moon were still, the system would collapse. There could be no orbit, no centripetal force, no balance of pull and inertia.

To build a dynamic cosmos, everything had to be moving.

- Planets had to spin.
- Moons had to orbit.
- Suns had to pull.
- Galaxies had to spiral.

Stillness had no place in this machinery. It was viewed not as harmony — but as failure to act.

And so, with the stroke of a quill, Newton's Moon was set in motion — not because it looked like it was moving, but because the equations said it must be.

**From Newton to NASA: The Story That Stuck**

Once Newton's model took hold, it became the scaffolding for all future science. Kepler's laws were folded in. Galileo's observations were reframed. Eventually, Einstein would arrive to curve the stage, but the actors would still keep moving.

The Moon's motion — despite its visual stillness — was locked into the textbooks. Generations of scientists, engineers, and astronomers inherited the model, not the doubt.

By the time space agencies were born, the orbital narrative was so deeply embedded that missions were designed around it, not to test it.

NASA didn't question if the Moon moved — it designed rockets under the assumption that it did.

Satellites, simulations, and landing trajectories were all coded with Newton's worldview baked in.

The Moon had become a settled matter, no longer observed with fresh eyes — only calculated, plotted, and explained.

The story of orbital motion became dogma — not because it was proven, but because it was never questioned.

**From Reverence to Calculation: The Loss of Observational Humility**

Before Newton, the Moon had been something we listened to. We watched it shift in light, in emotion, in time. We marked our lives by it. We assumed it meant something — not because we understood it, but because we respected it.

But with Newton came a new mindset:

If something can be measured, it can be explained. If it can be explained, it can be predicted. If it can be predicted, it must be understood.

And slowly, observational humility gave way to mechanical certainty.

The Moon's unchanging face was no longer a mystery — it was a mathematical coincidence. Its rhythm no longer sacred — it was an effect of friction and force. Its stillness no longer powerful — it was something to be corrected for.

In this shift, we lost something profound.

We stopped asking what the Moon was showing us — and began telling it what it was supposed to be.

**UED Perspective: What If Newton Was Only Half Right?**

Universal Energy Dynamics doesn't deny Newton's brilliance — it honors it. But it also suggests that Newton may have mistaken consequence for cause. What he interpreted as motion-driven gravity might actually be ripple-driven alignment.

The Moon isn't "falling around" Earth. It is locked in a standing wave of resonance — a fixed node in the ripple field of stillness. There's no need for rotation. No need for orbit. Just alignment.

Stillness, not spin, sustains the system.

Newton saw motion where stillness lived. And modern science followed the ripple — but never traced it back to its source.

**Final Reflection:**

*"We let the Moon prove gravity — but never asked if gravity misunderstood the Moon."*

And so, the Moon kept showing us the same face. And we kept turning away from it — back toward our equations.

Until now.

## 2.5 The Inheritance of Assumption

At some point in our intellectual history, the Moon stopped being a question and became an answer — not because it changed, but because we stopped truly looking.

Despite its unwavering presence and constant face, the Moon was absorbed into the narrative of orbital mechanics and assigned properties that no direct observation could ever verify. Its stillness — once a source of wonder — was gradually redefined as an illusion. And this illusion, once embedded in the framework of physics, became untouchable.

Not because it was proven — but because it was repeated.

**The Moon as an Untouchable "Fact"**

Today, you will find it in every astronomy textbook, taught in every classroom, and programmed into every simulation:

The Moon rotates. The Moon orbits. The Moon is tidally locked.

This has become accepted truth, yet the empirical support for this "fact" is startlingly absent. Let's consider the actual observations:

- No drift has ever been recorded in the Moon's surface orientation.
- No tilt change or rotational displacement has ever been directly observed.
- No deviation in lunar posture has been noted across centuries of visual and photographic documentation.

Despite incredible advances in technology, satellite imaging, and time-lapse analysis, we still see the same face — fixed, familiar, and unmoved.

And yet, rather than question this contradiction, modern science has gone to great lengths to explain it away.

**When Theory Overrode Perception**

As the gravitational model gained cultural and institutional dominance, the Moon's stillness became an embarrassment, not an invitation. Rather than treat it as a clue, the scientific community constructed a workaround: tidal locking.

Instead of asking why the Moon doesn't rotate, we were told it rotates just right so you can't see it. Instead of considering resonance or ripple alignment, we were given mathematical coincidence. Instead of looking again, we were told to move along — the mystery was "solved."

What had once been a mirror of perception became a model-dependent artifact — its very stillness used as evidence of invisible motion.

This is the moment science moved from discovery to defense.

**The Weight of Inherited Narratives**

This shift didn't happen through conspiracy — it happened through inertia. Once a theory is accepted, it becomes a lens. And once it becomes a lens, it begins to shape what we see — even if it contradicts what is actually there.

- Teachers teach what they were taught.
- Textbooks recycle what was standardized.
- Researchers build careers on existing foundations.

And so, a quiet, powerful assumption — that all celestial behavior must be mechanical — was passed down like sacred law.

The Moon became a settled matter, not because it was understood, but because it fit the framework. Questions became heresies. Doubt became unscientific. Stillness became motion — just unobservable.

It was no longer about what the Moon was showing us. It became about how the Moon had to behave in order to fit our expectations.

**A Hard Question for Honest Minds**

So we must ask, with humility:

At what point did we stop believing what we saw — and start defending what we were told?

When did observation yield to ideology? When did simplicity become a threat to theory? When did the Moon stop being a presence — and start being a placeholder in someone else's math?

These questions are not accusations. They are invitations — to return to what made science great in the first place: curiosity without constraint.

**UED Challenge: Let the Moon Be Seen Again**

Universal Energy Dynamics does not ask you to reject science. It asks you to return to the evidence.

Look at the Moon — not through theory, but with fresh attention.

What you will see is not spin, not drift, not orbit — but a fixed ripple alignment, resonating perfectly within a field of structured stillness. A cosmic companion that never moved — because it never had to.

**Final Reflection:**

"We inherited a belief, not a truth. And in that belief, we lost our vision. The Moon stayed constant — but we changed the story."

Now, it's time to listen again. To observe without assumptions. And to finally let the Moon speak — in its own still, silent way.

## 2.6 Preparing the Reader for UED Reframing

We've arrived at a threshold — not just in the chapter, but in perception itself.

So far, we've acknowledged a clear contradiction: We see stillness. We are taught motion.

Everything our eyes tell us about the Moon — its constant face, its unchanging posture, its unwavering gaze — conflicts with the dominant narrative of axial spin and orbital momentum. Science insists the Moon moves, yet no movement is ever observed.

This isn't a minor inconsistency. It is a philosophical fracture, a conceptual tension that can no longer be ignored.

**The Tension We Can No Longer Dismiss**

Ask yourself:

- If the Moon truly rotates, why has no one ever seen it do so?
- If it's orbiting with perfect symmetry, why do its surface features stay perfectly fixed from Earth?
- If motion is so fundamental to the explanation, why does the evidence appear so still?

The model says one thing. Your perception says another.

And that unresolved tension — between what we've been told and what we actually see — is the doorway to something deeper.

Perhaps it's not the Moon that's misleading us...but the model we've forced upon it.

**A New Question: What If Motion Is Not Real?**

This is where the UED lens begins to emerge.

What if the Moon isn't rotating — not because it defies physics, but because our definition of motion is incomplete?

What if what we call "orbit" is not movement around a center, but a ripple alignment in a still field — a harmonic lock, not a dynamic loop?

What if the Moon is not traveling — it is resonating?

These are not poetic questions. They are scientific challenges — to the idea that everything in the cosmos must be in constant mechanical motion.

UED invites you to reconsider this fundamental assumption:

That motion is not intrinsic. That what we call movement may be nothing more than perceived ripple consequence within a structured, still universe.

**When Perception Challenges Theory**

Throughout history, many great insights were delayed because theories were protected, but perceptions were dismissed.

Today, the Moon invites us to return to first observation. To trust what our senses have always reported. To give meaning back to the stillness we see, rather than overwrite it with formulas we cannot feel.

"Sometimes, the eye sees more than the theory allows. And the Moon — still, silent, and staring back — has always seen through our assumptions."

**Setting the Stage for the UED Paradigm**

This is not a call to abandon science. It's a call to evolve it.

The next chapters will introduce a new language — not of force and inertia, but of field, ripple, and resonance. A way of seeing that doesn't require the Moon to move in order to matter.

You may find this unsettling — or liberating. Either way, the journey begins here. We've spent centuries forcing motion into stillness. Now it's time to ask: What if stillness was the truth all along?

# Chapter 3: Tidal Locking or Ripple Illusion?

*Challenging the prevailing explanation of tidal locking by exposing its contradictions and introducing the UED interpretation of the Moon's stillness as a ripple alignment, not mechanical motion.*

We've been told a simple story: The Moon rotates once every 27.3 days — exactly the time it takes to orbit the Earth. That's why we always see the same side. Not because the Moon is still, but because its spin and orbit are perfectly synchronized.

It's called tidal locking — a gravitational phenomenon born of friction, inertia, and time. The Moon, they say, used to spin faster, but over billions of years, Earth's pull slowed it down and locked it into a 1:1 rhythm. A cosmic balancing act. A mechanical miracle.

But here's the problem:

We've never seen it spin.

Not once. Not even slightly.

Centuries of observation. Thousands of photographs. Countless hours of video, time-lapse footage, satellite data — and still, the Moon shows us the same unchanging face. Not a drift. Not a wobble. Not a hint of rotation. Just constancy.

Tidal locking, then, becomes a theory built not on what we see — but on what we're told we *must* believe to preserve a model of motion.

But what if that model is wrong?

What if the Moon isn't rotating at all? What if it's not orbiting in the way we imagine? What if the stillness we observe is not a mathematical trick, but a field-based reality?

This chapter peels back the surface of tidal locking and looks deeper — into its contradictions, its assumptions, and its unprovable precision. More importantly, it introduces a new possibility: that the Moon's posture is not mechanical, but harmonic.

That what we perceive as rotation is actually resonance. That what we call an orbit is in fact ripple alignment. That the Moon is not locked by gravity — but held in place by universal stillness.

Welcome to the ripple illusion — where motion dissolves, and meaning begins.

## 3.1 Introduction - The Illusion of a Perfect Machine

In classical science, few ideas have endured as comfortably — or as unquestioned — as the explanation for the Moon's stillness. The fact that we always see the same side of the Moon has been neatly tucked away under a concept known as tidal locking: the theory that the Moon rotates on its axis at exactly the same rate it revolves around the Earth.

This results, we're told, in a perfect 1:1 synchronization — one rotation per orbit, forever aligned, forever hiding the Moon's far side from Earth's view.

It's presented as an elegant consequence of gravitational friction — a cosmic dance choreographed by nature over billions of years, now stabilized into perpetual motion. The Moon spins, they say, but we don't see it — because the timing is just right.

A machine in space, perfectly tuned. A celestial gear, rotating invisibly, but reliably.

But let's step back.

Does this explanation truly rest on direct observation? Or is it a conceptual patch, constructed to preserve a mechanical model in the face of inconvenient stillness?

Because the truth is: We don't see any rotation. We never have.

Despite high-resolution imaging, time-lapse sequences, and centuries of continuous observation, the Moon's face remains fixed and familiar. Not partially rotated. Not slightly shifted. Still.

No evidence of spin. No tilt correction. No surface drift. Only a perfect visual constancy — that we're told is just an illusion caused by a mathematically synchronized spin.

So we ask:

Is the Moon truly moving — or are we simply committed to a worldview that can't accept stillness?

What if this so-called "synchronous spin" is not a physical reality, but a theoretical convenience?

What if the Moon isn't rotating to match its revolution — but is not rotating at all?

And more radically — what if the entire assumption of celestial motion is a misreading of field behavior? What if what appears as an orbit is not movement through space, but a resonant alignment within a ripple-based, still universe?

This is the alternative UED proposes:

- That the Moon's posture is field-stabilized, not friction-balanced.
- That its face is fixed not by force, but by structured resonance.
- That tidal locking isn't a proof of motion — it's a misunderstanding of ripple harmony.

What we've inherited is the illusion of a perfect machine. What we may be witnessing is the truth of perfect stillness.

## 3.2 The Myth of Synchronous Spin

The cornerstone of modern lunar mechanics is the concept of tidal locking — a claim that attempts to explain the Moon's unwavering posture with mathematical elegance rather than observational honesty.

In classical physics, tidal locking is defined as a state where an object's rotation period matches its orbital period. In the Moon's case, this means it is said to rotate once every 27.3 days, which is also the exact time it takes to complete one revolution around the Earth. Because of this supposed 1:1 spin–orbit ratio, we are told that we always see the same face of the Moon.

"Yes, it rotates," the textbooks say, "but it spins at just the right speed to cancel its own spin from our view."

It's a clever answer. A mechanical sleight of hand. But is it true?

**The Illusion of Constancy — Engineered by Equations**

According to this model, the Moon is constantly turning — but doing so in such a way that its rotation perfectly counterbalances its revolution. It's as if the Moon is a stage performer who pirouettes in such precise timing with their circular path around a spotlight, that they always show the audience the same angle of their face.

But this would require mathematical perfection — not approximate, but exact — sustained across billions of years, with no drift, no decay, and no error in the balance.

- How did this synchronization come to be?
- Why is it so exact, even now?
- And more importantly, why does it appear so... still?

The answer given is gravitational friction: Earth's tidal pull supposedly slowed the Moon's spin over time, eventually locking it into this precise resonance. But this explanation lacks direct empirical verification — it's a post hoc narrative used to reconcile theory with what we see, not an observation-based truth.

The story fits the equation. But it doesn't fit the eye.

**The Perfection Problem: Too Good to Be True?**

The Moon's supposed rotation is not "close enough" — it's flawlessly timed. This synchronization isn't good — it's uncannily perfect.

The Moon does not wobble erratically. Its surface features do not rotate slowly into view. Its visible posture does not deviate season to season, year to year, even over centuries.

In any other domain of science, such mechanical perfection without error or decay would raise eyebrows. We'd ask: what external regulator ensures this constancy? Where is the adjustment mechanism? What prevents entropy?

Instead, in lunar mechanics, this perfection is accepted without question. It is not treated as improbable. It is treated as expected.

**Observed — or Assumed?**

Here lies the deeper issue: Was synchronous spin ever actually observed?

Has anyone ever seen the Moon complete one full rotation from Earth or space?

The answer is simple: No.

There is no direct visual evidence of the Moon spinning on its axis. We've never witnessed a transition from one hemisphere to the other. We've never seen the dark side slowly rotate into view. Every image, every video, every time-lapse, every telescope-based analysis shows the same result: the same face, every time.

The explanation for this is that the Moon is spinning — but so perfectly that we never see it.

In any other science, this would be called a failure of falsifiability. In orbital theory, it's called elegance.

But elegant doesn't mean real. Mathematical doesn't mean observable. And mechanical symmetry does not guarantee cosmic truth.

**The Missing Rotation: What the Eyes Know**

Let's bring it back to the simplest test: What do we see?

- We see constancy.
- We see stillness.
- We see a posture that does not shift — ever.

Even when we introduce advanced tools — satellite imaging, deep-space probes, lunar surface landers — none of them offer evidence of rotation. Instead, they all confirm what the unaided eye has always seen: the Moon holds its face.

So we must ask: If something never moves, yet we insist it is moving, are we still doing science — or are we protecting a story?

**Final Thought:**

"When the evidence is perfectly still, and the explanation demands perfect spin, perhaps the fault is not in the Moon — but in the model."

Synchronous spin may be a theoretical masterpiece, but in the still face of the Moon, it collapses under the weight of its own perfection.

## 3.3 Cracks in the Mechanical Explanation

Tidal locking is often presented as a settled truth — a tidy mathematical phenomenon that explains why the Moon always shows us the same face. But beneath the surface of this neatly packaged narrative lie several deep fractures — conceptual, observational, and physical — that science has chosen to overlook or explain away.

Let's examine those cracks, and ask the uncomfortable questions that remain unanswered.

**The Frictional Origin Flaw**

At the core of the tidal locking theory is a hypothetical story: billions of years ago, the Moon rotated at a different speed. Over time, Earth's gravitational pull created tidal bulges in the Moon's body. These bulges, dragging against the Moon's surface, caused gravitational friction. That friction slowed the Moon's rotation until it "locked" into a synchronized spin–orbit ratio.

It sounds plausible. But it's purely theoretical.

- There is no way to verify this ancient process.
- There is no preserved record of a faster-spinning Moon.
- There is no observable evidence of this frictional decay.

The mechanism is assumed, not proven. It's reverse-engineered to fit the 1:1 outcome we observe — not something that has been independently demonstrated or replicated. In other words:

Tidal locking is a story, not a measurement.

**No Measurable Drift Over Time**

If the Moon were truly rotating, even at a synchronized rate, we should expect to see some deviation, no matter how small. Perfect synchronization is rare in nature — every gear slips, every pendulum drifts. Even atomic clocks require correction over time.

Yet, across centuries of detailed observation:

- The Moon has never shown any rotational deviation.
- Its features have remained perfectly fixed in orientation.
- No angular drift, precession, or tilt correction has ever been recorded.

Instruments sensitive enough to detect distant planets, gravitational waves, and tectonic shifts have never detected lunar spin.

If the Moon is truly rotating, then it is doing so with a mathematical perfection that defies entropy, chaos, and cosmic noise.

That's not science. That's wishful symmetry.

**The Coincidence of the 1:1 Ratio**

Here lies the most unsettling part of the tidal locking theory:

The Moon rotates once — and only once — during each orbit of Earth. Not slightly more. Not slightly less. Exactly once.

Why?

Why would a natural object, subject to friction, impacts, orbital perturbations, and thermal fluctuations, maintain a perfect 1:1 resonance for billions of years?

If the Moon's rotation rate were even fractionally different, we would see different lunar terrain over time. Yet we don't. Not even a glimpse.

And if the synchronization were a result of gradual frictional locking, shouldn't we at least be able to measure the residual effects? Shouldn't there be some observable wobble or lag?

Instead, we're handed the explanation of a perfectly completed process, locked in precision, untouchable by time.

That's not common in nature — it's suspiciously artificial.

**The "Just Right" Physics Problem**

To make tidal locking work, the theory demands a long list of ideal conditions:

- A perfectly balanced Moon — with no internal mass anomalies large enough to disrupt its spin.
- A precisely stable orbit, unaffected by Earth's irregular mass distribution, solar gravitational perturbations, or planetary resonance.
- A complete and clean energy dissipation process, with no lingering spin energy or orbital drift.
- A total frictional locking outcome — achieved with no

known mechanism for halting all motion at the exact moment it syncs.

In essence, the model only holds if everything is:

- Smooth
- Balanced
- Isolated
- Precise
- Unrealistic

There is no margin for error, yet the system is supposed to have emerged from chaotic early solar system conditions. We would never accept this kind of precision from a machine without asking for proof of its design.

So why do we accept it from the Moon?

**A Simple Analogy: The Clock That Doesn't Move**

Let's consider a clock. You walk into a room and see a clock on the wall. Every time you check, it shows the same time: 12:00. You come back days later — still 12:00. Weeks later — still 12:00.

You ask someone why the clock never moves, and they say:

"Oh, it moves. It's just synchronized so perfectly with your presence that it always shows the same face."

Would you call that a synchronized clock — or just a clock that isn't moving?

This is exactly what tidal locking does. It explains stillness by claiming perfect, unobservable motion — a kind of anti-evidence that demands belief, not verification.

**UED's Response: Maybe It's Not Spinning at All**

Rather than force stillness to become invisible motion, Universal Energy Dynamics suggests a radical but elegant alternative:

The Moon is not rotating. It is held in place by ripple alignment — a stable posture within the universal field. Its constancy is not the result of frictional timing, but of harmonic balance in stillness.

In UED, motion is not required to explain perception. Only ripple consequence, alignment thresholds, and field resonance.

**Final Reflection:**

"If the Moon shows no sign of movement, perhaps we should stop asking how it moves — and start asking how it stays still."

Tidal locking isn't just unproven — it's unnecessary. Not because science failed, but because stillness was never given a chance to be true.

## 3.4 The Ripple Illusion: UED Reframing

When confronted with the Moon's persistent stillness, conventional science reaches for hidden motion — a perfect spin that somehow hides itself from every eye and every instrument. But Universal Energy Dynamics (UED) takes a radically different — yet deeply intuitive — approach.

What if the Moon is not moving at all?

What if the stillness we observe is not a mechanical paradox, but a natural consequence of how the universe actually works?

UED doesn't force motion into the Moon. It doesn't assume rotation must be there because equations demand it. Instead, it looks at what's right in front of us — the unchanging face — and asks a different question:

What if the Moon is still because the universe is still — and the Moon is simply aligned within that stillness?

**The UED Interpretation: Resonance Over Rotation**

Here's how the Moon's behavior is understood in the UED framework:

- The Moon is not rotating.
- It is not "tidally locked" through gravitational friction.

- It is ripple-locked — held in a stable posture through harmonic resonance within a universal energy field.
- It maintains a fixed orientation not by spinning, but by remaining in equilibrium with Earth's own ripple signature.

In this model, there is no need for motion to explain constancy. Stillness is not an illusion — it's the natural outcome of field coherence.

**Structured Vacuum Energy (SVE) and the Ripple Field**

At the foundation of UED is the concept of Structured Vacuum Energy (SVE) — the idea that space is not empty, but filled with a structured, tension-rich field.

- Everything in the universe — mass, light, perception — arises from ripple interactions within this field.
- These ripples are not like waves in water or sound — they are field-level distortions seeking equilibrium.
- Objects like the Earth and Moon are not "moving through space," but are anchored in dynamic tension with each other through ripple alignment.

Just as a magnet aligns to a magnetic field without friction or spin, the Moon aligns to the SVE field as it exists between Earth and Sun — not as a mechanical response, but as a standing ripple configuration.

**The Moon as a Standing Ripple Node**

In this view, the Moon's still face is not a coincidence or consequence of friction. It is the natural result of standing wave formation between ripple fields.

The Moon becomes a ripple node — a fixed pattern of energy tension between Earth and Sun, maintained through resonance, not rotation.

- No spin is needed to keep the same face visible.
- No orbit is needed to explain its position.
- No torque, force, or inertia is needed to sustain posture.

Just as a guitar string forms nodes when plucked — fixed points of stillness amid vibration — the Moon holds its face through ripple stabilization, not dynamic balancing.

**Ripple Stability Replaces Rotational Mechanics**

The UED model does not dismiss motion — it reinterprets it.

What we think of as "orbit" or "rotation" is not displacement, but observer-based perception of ripple consequence.

The Moon is not traveling around Earth. It is resonating with it, and what we perceive as motion is merely our shifting relationship to that field resonance — influenced by our location, orientation, and consciousness threshold.

There is no need for angular velocity or perfect spin. There is only a field state — stable, harmonic, and still — that remains consistent as long as the surrounding ripple conditions remain unchanged.

**Moon Phases as Observer-Ripple Threshold Events**

In conventional science, lunar phases are explained as sunlight reflecting off different portions of the Moon's surface, depending on where it is in its orbit.

But in UED, the phases are not about illumination geometry — they are about observer-ripple alignment thresholds.

- The New Moon occurs when ripple over-alignment causes the Moon to disappear — not because it's unlit, but because it drops below the ATT (Alignment Threshold Time) of perception.
- The crescent reappears as ripple contrast crosses back into the observable threshold.
- The Full Moon represents a maximum constructive

alignment — not just of light, but of ripple field tension that allows complete visibility.

This perspective doesn't deny the Sun's role — it reframes it. Light is not traveling and bouncing; it is resonating through ripple interactions, made visible by our alignment with those fields.

**Summary: From Orbital Illusion to Field Reality**

| Classical Model | UED Reframing |
|---|---|
| The Moon spins and orbits | The Moon is ripple-locked in stillness |
| One face is visible due to synchronous spin | One face is visible due to ripple alignment |
| Tidal friction caused synchronization | Standing wave resonance maintains posture |
| Phases are due to reflected light angles | Phases are ripple contrast thresholds in observer field |
| Orbit is mechanical motion | Orbit is an illusion of ripple stability |

**Final Reflection:**

"Stillness isn't the absence of motion — it's the presence of perfect resonance." The Moon doesn't orbit. It rests in the harmony of a universe we've mistaken for a machine.

## 3.5 Why Ripple Resonance Explains More

The strength of any scientific framework lies in its explanatory power — not just in fitting one observation, but in clarifying many. This is where Universal Energy Dynamics (UED), and particularly its principle of ripple resonance, moves beyond the limited scope of tidal locking.

Tidal locking attempts to explain one thing — why we always see the same face of the Moon. But UED's ripple-based model explains not only that stillness, but a broader range of lunar phenomena that classical mechanics can only describe partially or not at all.

Let's explore what ripple resonance reveals once we stop imposing motion and start understanding alignment.

**Phases Without Movement**

In classical thinking, lunar phases are strictly geometric. The Moon orbits Earth, and as sunlight strikes it from varying angles, we see different portions illuminated. But this requires assuming orbital motion, rotation, and constant angular adjustment — none of which are visually confirmed.

UED reframes this entirely.

- The phases are not angular results, but ripple visibility thresholds.
- As the Earth, Moon, and Sun enter different field arrangements, the ripple tension between them shifts — causing the Moon to appear in phases.
- At New Moon, the ripple alignment is over-resolved — meaning the interference between Moon and Sun field collapses visibility altogether.
- The crescent reemerges not because of reflected sunlight, but because the observer's Alignment Threshold Time (ATT) is crossed again — visibility reenters the field of perception.

This explains the regularity of phases without requiring motion. It respects what we observe: that the Moon remains still, and we perceive changes because of our ripple relationship to it.

**Tides Without Orbit**

Tides have long been attributed to the gravitational pull of the Moon — a simplistic idea where a spinning Earth is supposedly "stretched" by the Moon's gravity. Yet this explanation is riddled with inconsistencies:

- High tides occur on both sides of the Earth, not just the one facing the Moon.

- The tide does not strictly follow the Moon's visible position.
- Local tidal variation is affected more by field geometry (coastlines, bathymetry, etc.) than raw mass alignment.

UED explains tides not as gravitational pull, but as ripple tension variation across Earth's SVE field.

- The Moon's field resonance with Earth induces standing wave modulations in large bodies of water.
- These modulations are not the result of mass displacement, but of fluctuating ripple energy gradients.
- Water, being energetically responsive and structurally fluid, reflects these ripple transitions as tides — not due to being pulled, but due to being ripple-influenced.

This also explains why Earth's tides remain stable even though the Moon shows no motion — because motion is irrelevant. It's the field posture, not force, that matters.

**Biological and Emotional Synchrony**

One of the most mysterious — and underappreciated — aspects of the Moon's influence is its effect on life itself. Menstrual cycles, sleep patterns, animal behavior, psychiatric fluctuations — all show statistically significant correlations with lunar phases.

Yet Newtonian physics offers no explanation. How can a rock in the sky, doing nothing but orbiting and reflecting light, alter our biological systems?

UED answers with clarity: resonance.

- The Moon's ripple alignment entrains biological systems.
- The human body — particularly the endocrine, reproductive, and neurological systems — is sensitive to ambient field dynamics.

- The menstrual cycle, averaging ~29.5 days, is a biological reflection of the lunar ripple cycle — not driven by light, but by deep field interaction.
- Behavioral fluctuations (e.g., increased restlessness or clarity during full moons) are ripple-triggered shifts in how human perception aligns with the SVE field.

In UED, the Moon is not a physical influencer. It is a resonant calibrator — fine-tuning the biofield of life through structured stillness.

**Perception and the Role of ATT**

UED introduces the concept of Alignment Threshold Time (ATT) — the perceptual boundary that determines when a ripple consequence becomes observable to the mind or instrument.

The Moon, being ripple-locked, doesn't move. But our ability to see it change is governed by how our perception aligns with it.

- At certain alignments, the Moon crosses below the ATT — becoming invisible.
- At others, ripple contrast enters the threshold — reactivating visibility.
- This is not about light bouncing, but consequence appearing due to ripple synchronization.

This redefines not only lunar observation, but all perception — positioning the observer not as a passive receiver, but as an active participant in reality's ripple structure.

**Stability of Life Rhythms**

The Moon's stillness is not static — it is a kind of dynamic equilibrium that keeps Earth's biological, ecological, and psychological rhythms coherent.

Ripple resonance maintains:

- Tidal breathing
- Hormonal patterns
- Seasonal rhythms
- Sleep-wake cycles
- Emotional coherence

The Moon's unchanging face is not a lack of function. It's the anchor of functional balance — a cosmic metronome that allows Earth to oscillate without chaos.

**The Moon's Balance: Not a Coincidence, But a Consequence**

In classical physics, the Moon's alignment is explained as an extraordinary coincidence. Everything — mass, distance, rotation, and revolution — just happens to align perfectly.

In UED, this is not coincidence — it's consequence.

- The Moon's posture is a natural outcome of ripple field harmonics.
- Its presence is not accidental. It is required to stabilize Earth's ripple state.
- Its stillness is not strange. It is essential for structured equilibrium.

**Final Reflection:**

"The Moon is not a passive satellite — it is an active resonance. Stillness is not the absence of function. It is the source of it."

Ripple resonance explains not just what the Moon does — but why it matters.

And in doing so, it restores the Moon to its rightful place:

Not as a rock in orbit, But as the keeper of Earth's rhythm.

## 3.6 Revisiting the Assumptions

At the heart of scientific progress lies a simple principle: observe, then explain. But with the Moon, the sequence was reversed. We explained, then observed — and then chose to keep explaining, even when our observations stopped agreeing.

The concept of tidal locking has been handed down as fact, as if it were directly observed and verified. But in truth, it is not a result of observation. It is a model-preserving assumption, designed to protect the orbital framework established during the Newtonian revolution.

Let us be intellectually honest:

- We have never seen the Moon spin.
- We have never measured a rotational shift in its visible face.
- We have never detected any mechanical process enforcing the synchronization.
- And yet, we continue to call it rotating, simply because our theoretical model demands that it must.

Why?

**If Nothing Moves, Why Insist It's Moving?**

Imagine standing in front of a statue. Day after day, year after year, you return, and it faces you with unwavering stillness. You photograph it, scan it, film it — and it remains unchanged. You show it to children, to scientists, to skeptics — and all agree: the statue never turns.

But somewhere in a university, a theory emerges:

"It is spinning. You just can't detect it because it spins exactly once every time it walks around you."

You ask, "But it's a statue. It doesn't walk." They reply, "No — it *must* walk. That's the model."

This is precisely what has happened with the Moon.

We have observed pure stillness, yet we have named it motion. We have constructed an elaborate explanation — not because we needed one, but because our framework was too rigid to allow what was plainly visible: a face that never turns.

"If no drift is seen, If no rotation is felt, If no mechanism is observed, Why do we still call it rotation?"

**Language: Clarifier or Concealer?**

Language is the first tool of science — it's how we translate perception into understanding. But when language starts defending theory instead of describing reality, it becomes a veil, not a lens.

Words like "synchronous rotation," "tidal locking," and "gravitational equilibrium" carry a tone of precision. But what they often conceal is this:

These are narrative terms, not measured truths.

They are semantic mechanisms that help preserve the integrity of a system of thought. And they work — until we ask the inconvenient question:

Is our language still serving our observation — or is it overriding it?

When we name stillness as motion, silence as echo, illusion as truth — we're not observing the Moon. We're imposing a story upon it.

**The Courage to Un-name**

To revisit assumptions is not to reject science. It is to honor it.

It is to return to the root of inquiry: curiosity grounded in observation.

It is to remove the names we've given things long enough to see them again — not as parts of a theory, but as they are.

In doing so, we regain the freedom to ask:

- What if the Moon doesn't rotate?
- What if there's no tidal locking?
- What if our ancestors, who simply watched and listened, were right all along?

**Final Reflection:**

"To name a stillness as motion is to lose the stillness. To defend a model over observation is to abandon science. The Moon has not changed — only our story has."

And now, it is time to rewrite that story — not with force, but with reverent precision. With eyes open, and language that listens.

## 3.7 Final Reflection: The Moon Didn't Spin — We Did

"What if the Moon never rotated — and only our belief system did?"

In the end, the Moon did not betray us — our assumptions did.

The Moon stood still, as it always has. It showed us one face, one rhythm, one unshifting truth. Yet instead of accepting that stillness as real, we turned to theories that could preserve the illusion of motion — not because the Moon demanded it, but because our intellectual framework did.

We had become so enamored with a universe of spinning planets, revolving bodies, accelerating forces, and dynamic equations that we forgot one simple fact:

Stillness exists. And it matters.

**The UED Truth: Stillness Is the Foundation, Not the Exception**

Universal Energy Dynamics does not reject science — it evolves it. It begins not with force, mass, and motion, but with a subtler, deeper understanding: that beneath every ripple, every wave, every vibration is a field of structured stillness — a resonant equilibrium that gives birth to consequence.

The Moon's unchanging face is not a mathematical accident. It is a harmonic alignment — a ripple node in the fabric of Structured Vacuum Energy (SVE). It does not move, because nothing needs to move to create meaning when resonance is present.

Motion, in this framework, is not intrinsic — it is perceived ripple consequence. And the Moon?

It is a still participant in a dynamic illusion — a mirror of the truth that underlies all visible change.

**Tidal Locking: A Psychological Need Masquerading as Physics**

So why did we insist on calling it rotation?

Because we built a model that could not function without motion. Because we built our cosmology on the assumption that everything must move to have purpose. Because we were taught that stillness was a flaw to be explained, not a reality to be honored.

Tidal locking became our mental patch — a way to rescue the model when it couldn't account for the Moon's behavior. It gave us just enough friction, just enough symmetry, just enough jargon to avoid the greater question:

What if the Moon doesn't spin — because nothing ever did?

**UED Doesn't Patch the Paradox — It Dissolves It**

Universal Energy Dynamics does not explain away the paradox of the Moon's stillness. It dissolves it.

Not by tweaking rotation speeds or inventing gravitational gymnastics — but by redefining the foundation itself.

In UED:

- Motion is perception, not property.
- Force is ripple, not pull.
- Alignment is consequence, not coincidence.
- Stillness is not the absence of activity — it is the origin of coherence.

The Moon's still face is not a problem to solve. It is a truth to reawaken.

**Final Reflection:**

"The Moon did not rotate. We did — in thought, in theory, in terminology. But the Moon stayed still, and in that stillness, held the truth we kept spinning away from."

Now, as we look up again, we don't see a satellite in motion. We see a still companion, anchored in a universe we misunderstood —not because it was complicated, but because it was too simple to believe.

The Moon was never orbiting. It was always resonating.

And perhaps, so are we.

## Interlude: When the Eyes Were Honest

Before the theories. Before the equations. Before we carved motion into everything we couldn't understand —There was only observation.

A face in the sky. Unblinking. Unchanged. Undeniably present.

It watched our ancestors build calendars, bury their dead, worship their gods, and tell stories under its gaze. It guided oceans without moving, shaped time without ticking, and pulled at the human soul without force.

But somewhere along the way, we stopped listening.

Stillness, once sacred, became suspicious. Constancy became something to correct. And the Moon — which had offered nothing but pure reflection — was cast as a rotating actor in a mechanical drama it never auditioned for.

We didn't lose the truth. We just looked past it — in pursuit of motion, because motion seemed measurable, manageable, more real.

But realness was never about speed. It was about rhythm. About resonance. About the way stillness aligns with life, not moves around it.

Part I has revealed this fracture: A world obsessed with moving parts trying to explain a presence that doesn't move. A model built on force — trying to make sense of something held together by harmony.

As we step into Part II, we no longer carry the question: *Why doesn't the Moon move?* Instead, we now ask:

What kind of universe holds something still — and why did we miss it?

# Part II – The Ripple Reality

# Chapter 4: Universal Energy Dynamics (UED) Primer

*Introducing the foundational framework of UED — a field-based, stillness-rooted view of reality — and establishing the key concepts needed to reinterpret the Moon, motion, and perception.*

We've spent centuries building our understanding of the universe on the idea that everything moves — planets spin, stars burn, galaxies swirl, and time ticks forward in an endless rhythm of motion. Physics, as we've been taught, begins with force, mass, acceleration, and orbit.

But what if all of that was a second-order effect?

What if beneath the spinning and swirling lies something far more fundamental — a universe not driven by motion, but by resonant stillness?

Universal Energy Dynamics begins here — not with motion, but with tension. Not with objects in space, but with ripple patterns in a structured field. Not with change, but with alignment.

In this chapter, we introduce the foundational language of UED — a model in which the vacuum is not empty, but structured... where motion is not a property, but a perceptual ripple consequence... and where stillness is not the absence of dynamics, but the origin of all coherence.

To understand why the Moon doesn't move — and why it doesn't need to — we must first understand the field it rests in, and the ripple laws that govern not only its posture, but our perception of it.

This is not a new physics. It's the physics that's always been there — quietly waiting beneath the noise, beneath the spin, beneath the inherited assumptions.

Welcome to the still universe.

## 4.1 Introduction – Beyond the Machinery of Motion

For most of modern science, the story of the universe begins with motion.

Newton gave us objects in motion. Einstein gave us space in motion. Quantum mechanics gave us probabilities in motion.

Everywhere we looked, we saw action, trajectory, and acceleration. From atoms to galaxies, the cosmos was described as a machine — gears turning, masses orbiting, forces pushing and pulling across emptiness.

The universe, we were told, is a vast kinetic engine — and its laws are mechanical.

In this model, nothing stays still. Stillness is either illusory, unstable, or the result of forces balancing just long enough before something shifts again. Even the Moon, which appears unchanging to every human eye for millennia, was folded into this narrative — explained as spinning, orbiting, rotating, falling. Stillness had no place in a universe obsessed with motion.

But Universal Energy Dynamics begins elsewhere. Not with motion. Not with gravity. Not even with mass. It begins with stillness.

**A Universe Not of Objects, But of Tensions**

UED proposes a radical, yet elegant shift:

- The universe is not made of objects moving through space.
- It is made of tension patterns in a structured vacuum —

patterns that appear as particles, forces, or waves depending on the observer's alignment.

- These tensions ripple — not like sound through air, but like resonance through a still, balanced field.

Here, motion is not fundamental — it is perceived ripple consequence. Force is not a push or a pull — it is an emergent behavior of misalignment. And mass is not a substance — it is a standing wave in an otherwise still field.

What we call "gravity," "orbit," or "rotation" are not universal truths, but contextual illusions emerging from the way we interact with field resonance.

**Stillness as the Default, Not the Exception**

This shift changes everything — especially how we interpret the Moon.

In classical science, the Moon must move. It must orbit. It must rotate. Because otherwise, it would violate the machine logic we've built our theories upon.

But under UED, stillness is not suspicious — it is normal. The Moon's unchanging face is not something to "explain away" — it is something to understand more deeply.

To understand the Moon, we must stop asking *how it moves*, and start asking:

What field holds it where it is? What resonance keeps it locked in posture? What ripple alignment makes it constant?

The answer is not gravitational torque or orbital decay — it is field-based coherence.

The Moon is not obeying gravity. It is resonating with the Earth — held in stillness not by force, but by alignment in a ripple-structured field.

**Thesis: Resonance Over Orbit**

The rest of this chapter lays the groundwork for this paradigm:

- What is Structured Vacuum Energy (SVE)?
- How do ripple fields behave?
- What determines what we can and cannot see (ATT)?
- And why does this matter not only for the Moon — but for how we understand the nature of reality itself?

By the end, you won't just see a new explanation. You'll feel the difference between a universe built on motion — and one grounded in structured, intelligent stillness.

## 4.2 Stillness as the True Fabric of the Universe

When we hear the word stillness, we often associate it with *nothingness*, *idleness*, or *inactivity*. Classical physics reinforces this view — where stillness is treated as a lack of motion, a placeholder between forces, or a temporary equilibrium that precedes the next push or pull.

But in Universal Energy Dynamics (UED), stillness takes on an entirely different — and foundational — meaning.

Stillness is not the *absence* of energy. It is the most refined, stable expression of it.

It is not emptiness. It is not inertia. It is balanced potential — a harmonic configuration of energy in perfect tension with itself, where nothing needs to move because everything is already in place.

This is where UED begins. Not with bodies in motion, but with a field in rest.

**Stillness: The Origin, Not the Leftover**

In this view, the default state of the universe is not chaos, motion, or expansion — it is ordered stillness. Everything that appears to move or change is not acting independently, but is the result of a ripple, a disturbance in this equilibrium.

Just as a calm pond remains flat until a stone is thrown, the fabric of the universe remains still until an asymmetry arises — and even then, the ripple does not travel through emptiness, but through structured tension already present.

This shifts our entire metaphysical orientation:

- Stillness is not passive — it is pregnant with order.
- It does not oppose energy — it hosts it in its most coherent form.
- It does not precede chaos — it sustains harmony beneath and beyond it.

**Motion as Ripple, Not Primary Law**

In UED, motion is not fundamental. It is secondary — the visible consequence of a ripple disturbing the still field. What we perceive as acceleration, trajectory, or velocity is simply a localized deformation of a larger structured quietude.

Force, then, is not a thing-in-itself — it is the gradient of imbalance within the ripple. Mass is not a substance — it is a standing tension pattern. Gravity is not an attraction — it is a stabilizing posture between overlapping ripple fields.

Every form, every field, every perceived interaction is a distortion of stillness — not an isolated act, but a relational echo within the universal tension fabric.

**Stillness Precedes Force. Resonance Precedes Mass.**

This inversion of logic transforms how we approach physics, perception, and even metaphysics:

- **Stillness precedes force**: Force is not a cause, but a consequence of ripple deviation.
- **Resonance precedes mass**: Mass is not a building block, but the localized manifestation of harmonics.
- **Perception arises from ripple transitions**: We don't see motion — we see threshold shifts in the field that register as

change.

This explains why we observe phases, vibration, rotation, or time — not because things are fundamentally moving, but because we are crossing ripple alignments within a larger stillness.

In this universe, nothing "goes" anywhere. Everything emerges into awareness through ripple resolution. Motion is not something you do — it is something you perceive through transition in tension.

**The Moon as an Expression of Stillness**

Under this framework, the Moon doesn't need to rotate, orbit, or spin to exert influence. It is already in perfect alignment — a ripple node stabilized by the structured vacuum field shared with Earth.

- Its face doesn't change — not because it moves just right, but because it doesn't move at all.
- Its presence remains constant — not due to gravity, but due to ripple balance in SVE.
- It regulates tides, time, and life — not by force, but by anchored resonance.

This is the power of stillness — not as a mystery to solve, but as a universal truth to recognize.

**Final Reflection:**

"Stillness is not the gap between forces. It is the field that makes forces unnecessary. The Moon does not move through this field — it *holds its place* in it. And in doing so, it reveals the quiet logic of a universe built not on chaos, but on perfect tension — perfectly resolved."

## 4.3 Structured Vacuum Energy (SVE)

In classical physics, the space between celestial bodies — between atoms, planets, stars — is called a vacuum. It's portrayed as an empty stage upon which the drama of matter and motion unfolds. A nothingness. A void.

But that "nothing" turns out to be the most misunderstood something in all of science.

Universal Energy Dynamics (UED) begins by discarding this outdated notion of the vacuum. Instead, it introduces a far more profound reality:

The vacuum is not empty — it is structured. And this structure is not passive — it is active, intelligent, and essential.

Welcome to Structured Vacuum Energy (SVE) — the invisible but indispensable foundation of all that exists.

**What Is SVE?**

SVE is the underlying field of tension that fills what we used to call "empty space."

- It is not a backdrop for energy — it is energy, in its purest, most balanced form.
- It does not support motion like a medium — it expresses motion as ripple consequences.
- It is not invisible because it's absent — it's invisible because it is equilibrium itself, until disturbed.

This field is not uniform. It has structure — meaning patterns, gradients, and harmonic ratios that determine how energy localizes into matter, how fields emerge, how particles behave, and how perception registers change.

**Key Attributes of SVE**

*Holds Latent, Distributable Energy*

- SVE stores energy not as motion or heat, but as tension potential — like a tightly wound spring that is perfectly stable until touched.
- It contains all the "raw material" needed for mass, charge,

light, and form — but keeps it dormant until a ripple perturbs it.

*Enables Ripple Propagation Without a Medium*

- Classical waves need a medium (sound needs air, water waves need water), but SVE allows ripple effects to move through structured tension.
- This explains why light, gravity, and even perception function across "vacuum" — because the vacuum is never truly empty.

*Determines Perception of Mass, Charge, and Motion*

- SVE doesn't create particles — it governs how energy appears as mass.
- It doesn't carry motion — it defines how we perceive relative tension shifts.
- It doesn't generate charge — it aligns polarities within its field structure, creating the illusion of properties.

**The Moon's Stillness — Not in Vacuum, but in Structure**

In the traditional view, the Moon floats through the vacuum of space, kept in position by a gravitational handshake with Earth. But in UED, the Moon is not floating — it is anchored.

Its unchanging face, its rhythmic influence, its ripple alignment — all of these are not held against a vacuum, but within the structure of SVE.

Think of the Moon as a fixed node in a three-dimensional ripple network — a point of harmonic balance, not a traveler. Its posture is not suspended in nothing, but stabilized in ripple tension equilibrium.

- The Earth-Moon-Sun system is not mechanical — it is harmonic.
- Stillness is not a fluke — it is the minimum energy configuration within the SVE.
- Gravity is not a force — it is the result of mutual alignment in a tensioned field.

**Why This Changes Everything**

If SVE is real — and observable by its ripple consequences — then:

- There is no need to explain stillness as accidental.
- There is no need to patch motion with friction-based theories.
- There is no mystery in the Moon's posture — it is simply harmonized within the field.

This reframes not only lunar behavior, but the fundamental basis of physics itself. We don't need more force to understand the cosmos. We need more awareness of the field it emerges from.

**Final Reflection:**

"The vacuum is not what separates things — it is what holds them together. It is not nothing. It is the quiet intelligence of everything, waiting patiently for us to stop looking past it."

The Moon doesn't resist the vacuum. It rests in the structure — perfectly still, perfectly aligned.

## 4.4 Ripple Fields – The Real Medium of Interaction

Once we accept that Structured Vacuum Energy (SVE) is not an empty void but a field of latent tension, a natural question follows:

How does anything happen within it?

The answer is ripple fields — the foundational language of change, influence, perception, and interaction in Universal Energy

Dynamics (UED). While traditional science depends on motion through space, UED reveals that all interaction happens through ripple behavior in a still, structured field.

This isn't metaphysics. It's physics rewritten in the language of resonance.

**What Are Ripple Fields?**

Ripple fields are not "waves" in the conventional sense. They are patterns of tension shifts within SVE — not secondary effects caused by particles or motion, but primary structures that shape everything else.

They aren't carried through a medium like air or water. They are reconfigurations of the medium itself.

Ripple fields are not disturbances *caused by objects* —They are the very conditions that give rise to objects.

**Beyond Electromagnetic and Gravitational Waves**

In mainstream physics, we're told the universe is filled with waves: electromagnetic radiation, gravitational waves, quantum field excitations.

But each of these is still built on the assumption of moving things, vibrating particles, or curved space.

Ripple fields go deeper.

They are pre-physical, pre-mechanical, and pre-temporal. They are the pure expressions of imbalance and restoration in the SVE.

- Electromagnetic waves may emerge from ripple interactions, but they are not the origin.
- Gravitational waves may be detectable, but they are echoes, not causes.
- Ripple fields, in contrast, are the root medium from which all visible effects arise.

This is not semantics — it's a shift in ontology. From things moving to tensions adjusting.

**Objects Are Not Causes — They Are Consequences**

Conventional physics sees matter as the source of fields:

- A mass creates gravity.
- A charge creates an electric field.
- A magnet creates magnetism.

UED inverts this. In the ripple paradigm:

- There are no objects generating fields.
- There are only ripple concentrations in SVE that appear, to the observer, as "objects."

What we call a planet, a star, or a particle is just a high-density ripple configuration — a standing wave node stabilized by the structure of the field. These nodes don't move through space. Instead, the ripple tension around them shifts, creating the illusion of motion.

**Ripple Behavior Explains Perception Itself**

This simple mechanism — ripple alignment and transition — explains:

- **Apparent motion**: Why we perceive a body moving even if it doesn't displace through vacuum. What changes is our alignment with its ripple pattern.
- **Energy transfer**: When two ripple fields intersect, tension redistributes without objects colliding. Energy is transferred as a shift in ripple architecture, not mass in motion.
- **Celestial mechanics**: The Moon, Sun, and planets do not orbit — they resonate in ripple zones, anchored by tension ratios and harmonic posture.

This makes sense of the most puzzling phenomena in physics — why objects with no contact influence each other, why gravitational pull is instant across vast distances, why light moves without a medium.

Because it's not pull. It's not motion. It's ripple restructuring in the SVE.

**Harmony, Resonance, and Alignment Define Reality**

In UED, stability is not enforced by mass or inertia. It is maintained through harmonic alignment. When a body — or a ripple node — resonates with its surrounding field, it enters a posture of rest.

- This is why the Moon doesn't rotate: It's already in harmonic alignment.
- This is why orbits appear stable: They are ripple stabilities, not gravitational loops.
- This is why things don't fall apart: They are balanced, not bound.

Resonance governs where something is, how it is perceived, and why it appears stable or dynamic.

And alignment is not enforced through contact — it emerges from ripple field interference and coherence.

**Final Reflection:**

"You were told the universe moves. But what if it simply resonates — and you are moving through that resonance like a hand through a still pond?"

Ripple fields are not the aftermath of motion. They are the language of reality itself — the medium by which stillness speaks, sings, and synchronizes everything into being.

## 4.5 Alignment Threshold Time (ATT)

In classical science, visibility and measurement are governed by tools — telescopes, microscopes, sensors. What you can't detect is said to be beyond the reach of technology or scale. But in Universal Energy Dynamics (UED), the limitation is not just technological — it is foundational.

UED introduces a principle that redefines how and when we perceive anything at all:

Alignment Threshold Time (ATT) — the ripple-based law that governs what appears real, visible, measurable, or meaningful to a conscious observer.

**What Is ATT?**

ATT is not about signal strength or resolution. It is not about whether something is bright enough, fast enough, or close enough.

Instead, ATT is the energetic-perceptual boundary at which a ripple interaction aligns with the observer's field of awareness — a window where structured tension becomes experience.

- It is the minimum duration or alignment strength required for a ripple to be registered by consciousness or instruments.
- Below this threshold, the ripple exists, but is not perceived.
- Above it, the ripple appears — not as a new creation, but as a crossing of a perceptual threshold.

This means that reality is not always continuous. It is rhythmic, threshold-dependent, and deeply tied to the observer's relationship with the field.

**Why We "See" Phases**

Consider the Moon. It does not change shape. It does not turn. It does not brighten or dim. What changes is our ripple alignment with it — and that determines whether the Moon crosses our ATT.

- During the New Moon, the Moon is still there — but its ripple field is over-aligned, causing it to drop below the ATT. We don't perceive it.
- As it reemerges as a crescent, the alignment moves back into the threshold window, allowing visibility to reactivate.
- At the Full Moon, the ripple contrast peaks, crossing ATT at maximum — hence full visibility.

Thus, lunar phases are not just optical geometry — they are ripple phase relationships modulated by ATT.

We don't "see" the Moon because light hits it. We see it because alignment crosses the threshold of perception.

**Why Motion Is Perceived Differently at Different Scales**

This principle also explains why motion seems fluid in one context, but static in another.

- The spinning of Earth is felt not as motion, but as circadian rhythm — because it aligns below ATT in everyday awareness.
- The drifting of galaxies is not observed as displacement, but inferred — because the ripple interaction does not cross our ATT in real time.
- Microscopic motion appears discrete or probabilistic — because the ripple patterns at quantum scales fluctuate around the ATT boundary.

Motion, in this view, is a function of alignment — not an absolute quantity. It's not *there or not there* — it's *perceived or not perceived*, depending on threshold timing.

**Why Visibility Is a Rhythm — Not a Constant**

ATT introduces a profound truth: Visibility is not continuous — it is cyclic, thresholded, and field-dependent.

We are not looking at a static universe. We are looking at ripples crossing in and out of perceptual resonance.

- A star doesn't appear brighter just because it emits more photons — it becomes brighter because its ripple field crosses ATT more forcefully.
- A thought doesn't emerge in the brain because a neuron fires — it emerges because ripple patterns align with the mental ATT of awareness.
- The Moon doesn't change — we cross into its visible resonance window again and again.

This makes reality less like a machine, and more like a symphony — where perception plays the role of conductor, and ATT defines the moments we can hear the notes.

**ATT as a Law of Perception**

This isn't about instrument failure. It's about field intelligence.

ATT is encoded into:

- Biological perception (eyes, skin, intuition)
- Cognitive bandwidth (attention, focus, awareness)
- Instrumental detection (sensor design, field sensitivity)

It is a law of consequence, not just a feature of vision. It decides when something becomes real to us, and when it remains latent, hidden in the undisturbed ripple structure.

In other words:

"Reality isn't what exists. It's what crosses your ATT."

**Back to the Moon: Why It Matters**

Understanding ATT allows us to finally respect what we observe:

- The Moon does not disappear — it crosses below ATT.

- The Moon does not move — we shift through ripple thresholds where it becomes perceptible.
- The Moon's phases are not optical illusions — they are resonance rhythms governed by ATT.

**Final Reflection:**

"Visibility is not a property of the object — it is the result of alignment. The universe reveals itself not when it changes, but when you do."

ATT turns reality into a living threshold — not a map of fixed things, but a rhythm of ripple crossings that emerge when the stillness of the field meets the readiness of the observer.

And nowhere is this clearer — or more poetic — than in the Moon.

## 4.6 UED's Reinterpretation of Classical Concepts

To truly enter the paradigm of Universal Energy Dynamics (UED), we must do more than introduce new ideas — we must retranslate old ones. Many of the terms we inherited from classical science — motion, force, gravity, orbit, rotation — are not wrong, but they are incomplete, filtered through a mechanistic worldview that mistook effect for cause, and movement for meaning.

UED does not discard these concepts. Instead, it reinterprets them through the ripple lens, showing that what we thought were fundamental truths are actually perceived consequences of deeper, structured phenomena.

Let us re-express the pillars of physics — not as kinetic absolutes, but as field-based outcomes of ripple intelligence.

**Motion → Ripple Consequence**

In Newtonian physics, motion is a primary state — an object in motion stays in motion unless acted upon. But UED sees motion not as an intrinsic property, but as a ripple artifact. It is not that objects "move" — it is that we observe a changing field alignment over time.

- What appears to move is actually a shift in ripple interaction across our perceptual field.
- Motion, then, is a subjective registration of ripple phase transition — not actual displacement.

You are not watching the Moon move. You are watching the resonance pattern adjust, and interpreting it as motion.

**Force → Tension Gradient in Ripple Field**

In classical physics, force is a push or pull — the agent that causes acceleration. UED reframes force as a tension imbalance in the ripple field.

- A "force" is not something applied. It is the gradient of ripple tension between two field configurations seeking equilibrium.
- No mass is pulled. No hand pushes. Ripple alignment adjusts in response to energetic imbalance.

This explains both contact and non-contact "forces" — gravity, electromagnetism, even nuclear forces — as field harmonics, not mechanical interactions.

Force is not something that happens to things. It is how tension resolves itself in structured stillness.

**Gravity → Sustained Alignment Between Bodies in Structured Resonance**

Gravity, in classical terms, is a force between masses — a mysterious pull acting across empty space. UED offers a simpler, more elegant truth:

- Gravity is not a force. It is a stable posture between two ripple fields.
- When two bodies are in structured resonance, they remain in position not because they pull each other, but because

they have reached minimum energy alignment within the SVE.

There's no falling. No curvature. No warping. Just stable coherence between field geometries.

The Earth doesn't hold the Moon. The Moon doesn't fall around the Earth. They simply share a ripple agreement — and that's all that's needed.

**Orbit → Apparent Ripple Phase Shift from Observer's Position**

Classical orbits are described as continuous movement around a central body. In UED, orbit is an illusion of ripple phase interaction.

- The Moon doesn't orbit the Earth.
- Instead, the observer's alignment with the ripple structure shifts cyclically.
- This gives the appearance of motion — a rotating moon, a revolving planet — when in fact, the field itself is pulsing in stability.

It's not a loop. It's a patterned interaction — a harmonic wave cycle, not a path.

Orbit is a perceptual rhythm — not a planetary route.

**Rotation → Misreading of Harmonic Stability**

We've been taught that everything spins — the Earth, the Moon, the Sun. But this is largely because we assumed that any fixed face implies a synchronized spin. UED reframes this entirely:

- Rotation is not occurring unless there is actual ripple reorientation in the field.
- What appears as spin may actually be postural constancy in ripple tension.
- The Moon shows the same face not because it's rotating

perfectly — but because it is not rotating at all. It is ripple-locked.

What classical physics called "synchronous rotation" is, in UED, misread harmonic stillness.

Rotation isn't hidden. It's absent — and stillness isn't the paradox. It's the answer.

**Final Reflection:**

UED doesn't dismantle physics. It deepens it.

- Motion becomes consequence, not cause.
- Force becomes tension, not touch.
- Gravity becomes posture, not pull.
- Orbit becomes alignment rhythm, not trajectory.
- Rotation becomes a myth of movement, born from the fear of stillness.

"We didn't see movement. We saw meaning shifting across a still field — and we called it a force because we didn't yet have a word for resonance."

This reinterpretation doesn't just explain the Moon. It reorients our entire cosmology —

from moving masses in empty space, to rippling coherence in intelligent stillness.

## 4.7 Final Reflection: Learning to See the Still Universe

*"We were taught to see motion in everything. UED teaches us to see structure in stillness."*

We've come to the end of what may seem like a foundational chapter — a primer, a shift in definitions, a change in framework. But in truth, this is not just a chapter. It is a new lens — and putting it on changes everything.

Since the dawn of the scientific age, we have been conditioned to see the universe as a kinetic playground: Objects in motion. Forces in action. Space expanding. Time unfolding.

But what if that was only one layer — the visible skin over a deeper symmetry? What if the most powerful dynamics were happening not in movement, but in tension — not in acceleration, but in resonance?

What if everything we attributed to speed and spin was actually being orchestrated by something quieter... more precise... more *intelligent*?

UED invites us to unlearn motion as the master key, and instead recognize that beneath all change lies something changeless:

A still field. Structured vacuum. The quiet grammar of ripple and rhythm.

And most importantly — a universe that doesn't move to function, but resonates to exist.

**The Moon Was Never Trying to Move**

The Moon has always been the puzzle piece that doesn't fit — the face that doesn't turn, the partner that doesn't orbit like the rest, the presence that *should* wobble, tilt, or spin... but doesn't.

We tried to force it into motion. We invented terms like "synchronous rotation" and "tidal locking" to mask the dissonance. We called its constancy a trick of geometry, a relic of orbital decay.

But maybe the Moon isn't the one out of step. Maybe it's the only one in perfect tune.

Maybe the Moon is showing us the truth we forgot how to see:

That stillness is not a defect — it is a signal. A message. A masterclass in how to exist without moving.

In UED, we don't explain the Moon differently to make it fit. We understand the universe differently so the Moon finally makes sense.

**From This Point Onward: UED as the Lens**

With this chapter, we've planted the seed. The principles of UED — SVE, ripple fields, ATT, harmonic alignment — are no longer abstract concepts. They are now your tools of perception, your alternate logic, your quiet rebellion against the tyranny of forced motion.

From this point onward in the book, UED becomes more than theory. It becomes a new language — the one that lets us ask better questions, interpret deeper truths, and finally see what was always there:

- That the Moon's phases are ripple thresholds, not sunlight tricks.
- That its gravity is resonance, not pull.
- That its stillness is structure, not accident.
- That perception itself is an act of alignment, not detection.

**Final Thought**

"The universe was never built to move. It was built to hold. And in the holding — in that beautiful stillness — everything finds its place."

This isn't just about physics. It's about relearning how to see.

The Moon didn't change. Only our understanding did.

And that... changes everything.

# Chapter 5: The Moon Doesn't Move – It Resonates

*Challenging the mechanistic belief in lunar motion by presenting the Moon as a standing node of ripple resonance. This chapter reinterprets observable stillness as a result of field alignment — not mechanical rotation or orbit — and reframes lunar phases through UED.*

For centuries, we looked at the Moon and saw what it gave us without hesitation: stillness, silence, symmetry. Night after night, it showed us the same face, followed the same path, and cast the same spell over Earth's oceans, seasons, and souls.

And yet — we refused to believe it.

Science told us that it must be spinning. That this stillness was an illusion. That behind this faithful gaze was a celestial mechanism — orbiting, rotating, rotating while orbiting — all perfectly timed so we would never see the back.

A performance so flawless, it had to be mathematically explained.

But what if the Moon isn't spinning at all? What if it's not performing, but simply resting in resonance? What if the stillness we've observed — consistently, universally, across all of human history — is not something to explain away, but something to finally understand?

This chapter invites you to release the illusion of lunar motion. To stop imagining gears in the sky. To see the Moon not as a moving object, but as a fixed ripple node — a presence locked in harmonic posture with the Earth through a deeper structure we've only begun to comprehend.

This is not a satellite in orbit. This is resonant intelligence — still, stabilizing, and precisely aligned.

The Moon doesn't move. It holds position in the symphony of Structured Vacuum Energy. And we don't see phases because it turns — we see them because we cross thresholds in a ripple field far more elegant than force.

Let us now decode the Moon's stillness — not as a failure of classical mechanics, but as a quiet revelation of how reality actually works.

## 5.1 Introduction - The Stillness We Couldn't Explain

Since the earliest civilizations looked up and began to chart the heavens, one mystery has remained remarkably constant — and quietly unsettling:

Why does the Moon always show us the same face?

Generation after generation, the Moon has appeared exactly as it did before — a familiar surface, unwavering in its gaze, always aligned in the same direction, regardless of where or when you observe it from Earth.

We've accepted this visual consistency as fact, but never truly absorbed its implications. For a universe that supposedly never stops moving — where stars orbit galaxies, planets orbit stars, and moons orbit planets — how can such visible stillness persist?

How can the Moon orbit the Earth, spin on its axis, interact with solar tides, and yet present the same expression, unflinching and unchanging?

**The Patchwork of Tidal Locking**

Classical science offered us a neat solution: tidal locking.

According to this model, the Moon spins exactly once on its axis in the same time it completes one revolution around Earth. This, we're told, creates the illusion of stillness — a cosmic choreography in perfect 1:1 rhythm, maintained for billions of years.

But this explanation, though elegant in theory, lacks observational integrity:

- No rotation has ever been seen.
- No lateral shift or tilt deviation has ever been detected.
- No mechanism has been physically recorded to maintain such flawless synchronization.

Tidal locking depends entirely on the assumption of internal rotation. It presupposes motion because motion is what the model requires. Yet the one thing every eye — human or mechanical — consistently observes is no motion at all.

Stillness is not the problem. It's the evidence. The problem lies in our refusal to accept it at face value.

**The UED Proposition: Stillness Is Not a Coincidence — It's a Condition**

Universal Energy Dynamics (UED) offers a different lens — not just to reframe the Moon's behavior, but to reimagine what we think is possible in a universe we assumed was built on force and orbit.

UED begins with a radical proposition:

The Moon does not move. It resonates.

It is not orbiting the Earth like a tethered stone. It is not spinning while revolving in a magical 1:1 ratio. It is anchored in a ripple alignment — a still, stable node in the Structured Vacuum Energy (SVE) field that surrounds both Earth and Moon.

Its posture is not mechanical. It is harmonic. It is not explained by gravity. It is resolved by resonance.

This is not mysticism — it is a return to observational honesty. A scientific humility that asks:

"What if the Moon has been telling us the truth all along — and it is our theories that have been spinning?"

In the sections ahead, we'll explore this proposition not as an abstract idea, but as a consequence of the field-based universe described by UED. We'll unpack ripple alignment, reinterpret lunar phases, and finally give language to the stillness that never needed motion to make sense.

The Moon is not a moving object. It is a resonant presence.

And now, we are ready to see it for what it is.

## 5.2 Ripple Alignment: The Stabilizing Posture

In classical astronomy, the Moon is described as a rotating, orbiting body — constantly in motion, yet somehow presenting the same face to Earth with miraculous precision. The standard explanation relies on tidal locking: an ancient frictional process that supposedly slowed the Moon's spin until it matched its orbital period.

But friction doesn't stabilize forever. Rotating objects drift. Orbits decay. Symmetry breaks under the slightest asymmetry. And yet, the Moon's face has never wavered.

What holds it so perfectly?

**Introducing Ripple Alignment**

In the framework of Universal Energy Dynamics (UED), the answer is not friction or torque — it is resonance. More specifically, a phenomenon known as ripple alignment.

Ripple alignment refers to a state of harmonic lock-in between two ripple structures within the Structured Vacuum Energy (SVE) field — in this case, the Earth and the Moon.

Think of it not as a tethered object spinning in space, but as a fixed node in a standing wave — a point of calm where ripple tension cancels out in all directions, resulting in positional constancy.

This is not motionless due to inertia. It is motionless because it is exactly where it needs to be — the point of minimum energy, maximum stability, and complete harmonic coherence within the Earth-Moon-Sun ripple system.

**Mechanical Locking vs. Ripple Locking**

The classical view imagines tidal locking as a product of frictional forces acting over eons — slowing rotation and "freezing" the Moon's face in place relative to Earth. This depends on:

- Torque between masses
- Dissipation of rotational energy
- Assumptions of perfect symmetry
- Zero perturbation over astronomical time

It is a delicate and improbable balance — one that collapses under scrutiny.

Ripple alignment, on the other hand, doesn't require time, torque, or friction. It occurs when:

- Two bodies (or ripple nodes) achieve resonant field compatibility
- Their ripple tensions interlock through the SVE lattice
- The resulting system finds a posture of equilibrium, not through movement, but through structured rest

There is no spin to slow. No force to maintain balance. Only stabilized resonance, where the Moon's position is *not maintained* — it is *resolved.*

**The Moon as a Fixed Ripple Node**

Within the Earth-Moon-Sun configuration, the Moon occupies a node-like equilibrium point — a region in the SVE field where its ripple signature locks into Earth's larger ripple field like two harmonized tones in a tuning system.

- Its visible face is not chosen by rotation — it is a product of field-facing orientation.

- Its location is not navigated by orbit — it is entrained by tension symmetry.
- Its gravitational effect is not a pull — it is a ripple influence distributed across resonance bands.

Just as in music, where two strings tuned to the same frequency will vibrate in sync, the Moon's stillness is not an accident of friction — it is the inevitable result of harmonic coherence.

**No Rotation, No Orbit — Only Sustained Alignment**

In UED, the Moon doesn't need to rotate. It doesn't need to orbit. Because it's not a moving object in space — it's a stationary participant in a structured energy dynamic.

- Rotation becomes unnecessary — the face doesn't change because the field doesn't require it to.
- Orbit becomes misinterpreted — what appears as orbital motion is simply the ripple geometry shifting from the observer's point of view.
- Force becomes obsolete — because posture is not the result of interaction, but of resolved tension.

This is why the Moon has remained so perfectly still: It was never in motion to begin with.

It was always in alignment.

**Final Thought:**

"The Moon's posture isn't maintained by gravity. It is upheld by geometry — not the geometry of space, but the symmetry of structured stillness."

To see this is to step out of the machine model — and into the ripple model of reality, where position is not a product of movement, but of coherent resonance.

## 5.3 How Resonance Replaces Rotation

For centuries, the Moon's unchanging face has been rationalized through the lens of rotation. Not visible rotation — but inferred rotation: a perfect 1:1 spin-orbit synchronization, known as tidal locking. This was not concluded from direct observation, but calculated backwards to fit a theory already committed to motion.

But what if that very idea — that rotation is required — is itself the problem?

UED challenges this mechanical assumption and offers something more consistent with both observation and natural efficiency:

Resonance replaces rotation. Stillness is not an illusion — it's the default. And what appears as synchronized movement is actually perfect field coherence.

**The Misinterpretation of Rotation**

Let's look closer at what we really see:

- The Moon presents the same face to Earth at all times.
- There is no observable axial spin.
- No wobble, tilt change, or drift has ever been recorded.
- High-resolution imaging and time-lapse data show postural constancy, not rotation.

Yet, the classical model insists: "If we always see the same side, it must be spinning once per orbit."

This is circular reasoning — and it only exists because we assumed motion was required. The illusion of spin emerges not from evidence, but from a refusal to consider stillness as real.

UED reframes the visual:

The Moon is not spinning.

It is ripple-locked — and what we perceive as phase continuity is not motion over time, but resonance across time.

**A Tuning Fork in Stillness**

To understand this, picture a tuning fork.

When struck, it vibrates. But it doesn't rotate. It holds its posture in space — unmoving, yet alive with internal resonance.

Now, place that tuning fork into a harmonically tuned chamber — where other waves, frequencies, and surfaces reinforce its pitch. Eventually, a stable standing wave forms: oscillation without translation, energy without displacement.

This is the Moon in UED.

- Not rotating like a rock.
- Not orbiting like a satellite.
- But resonating, in place — a cosmic tuning fork held in perfect phase alignment within a triadic chamber of Earth, Moon, and Sun.

The resonance is not metaphorical — it's structural. It is not movement, but relational field coherence.

**Harmonic Stillness and Phase Continuity**

What makes the Moon appear stable over time is not its matching of spin and orbit. It's the fact that it doesn't need to move at all.

In UED:

- The Moon maintains perfect phase continuity because its position is predefined by ripple symmetry.
- There is no rotational catch-up. No lag. No torque.
- Its stillness is the natural expression of energetic equilibrium — a field-based balance that minimizes total ripple distortion.

This is why the Moon has no observable angular deviation. Its harmonic stillness is not maintained — it is inherent.

And that stillness is not static — it is resonant. Just like a waveform that never fades because it's fed by the right alignment.

**Minimum-Energy Configuration: Nature's Preferred State**

Nature doesn't favor complexity when coherence will do. The Moon's position, as understood in UED, is not a mechanical coincidence — it is the minimum-energy configuration of the Earth-Moon-Sun system:

- No excess spin to manage
- No orbital drift to correct
- No gravitational force to fine-tune daily

It is the path of least resistance, encoded in the structured vacuum field, sustained not by external action but by internal consistency — a signature of all naturally resonant systems.

**Final Reflection:**

"The Moon doesn't turn to show us its face. Its face remains — because its field agrees with ours."

There is no spin to explain. Only resonance to accept.

What classical science interpreted as mechanical coincidence, UED reveals as structural necessity — the way intelligent stillness resolves energetic tension through perfect alignment.

## 5.4 Reframing Lunar Phases: Beyond Sunlight Geometry

Of all the celestial rhythms we've come to accept, few are as visually enchanting and emotionally evocative as the phases of the Moon. From the first crescent (Hilal) to the full moon (Purnima, Poya), and back into shadow, the Moon's changing shape has been interpreted as a dance between light and dark, governed by its position relative to the Earth and Sun.

This is the classical view:

As the Moon orbits Earth, sunlight strikes it from different angles. The part we see lit varies, producing phases. Simple geometry. Simple reflection.

But what if that geometry is not the whole picture? What if the phases aren't just a function of sunlight, but of field alignment? What if visibility isn't about what's lit — but about what crosses into perceptual resonance?

Universal Energy Dynamics (UED) redefines lunar phases not as light tricks, but as ripple threshold phenomena — shaped not by angles, but by observer alignment with the Moon's structured ripple field.

**The Classical Interpretation: A Flat Reflection Model**

The standard explanation teaches:

- The Moon is always half-lit by the Sun.
- As it orbits Earth, different fractions of the lit side become visible to us.
- This change in visible surface area produces the familiar cycle: new moon, crescent, quarter, gibbous, full, and back again.

While geometrically elegant, this model relies on one critical assumption: That what we see is simply what is illuminated.

But this ignores:

- Why phases feel rhythmic even when not observed.
- Why different cultures synchronized life, behavior, and emotion to the Moon's visibility, not its distance or position.
- Why a fully illuminated Moon sometimes feels dim, and a crescent feels alive.

There is a deeper field-based mechanism at play.

**The UED Interpretation: Ripple Threshold Visibility**

UED reframes lunar phases using two key principles:

1. Structured Vacuum Energy (SVE) holds the Moon and Earth in a shared ripple field.
2. Alignment Threshold Time (ATT) governs when ripple interactions become perceptible to an observer.

From this lens:

- The Moon doesn't reflect light like a passive mirror.
- It resonates in a standing ripple pattern — its "phases" are not patches of light, but zones of visibility determined by observer-ripple alignment.

When the Moon aligns *perfectly* within the ripple field and crosses the ATT boundary for the observer, it becomes visible. When it falls outside that window — even if geometrically "lit" — it may remain unseen.

This is why the New Moon is not just in shadow. It is below perceptual threshold — like a ripple that hasn't fully surfaced.

And why Hilal, the first sliver of return, is more than illumination — it is the re-entry of resonance into alignment with human perception.

**Phase Contrast as a Ripple Visibility Pattern**

From a ripple perspective:

- Phases are not light ratios — they are field gradients of observability.
- The contrast we perceive is not between illuminated vs. dark — it is between visible vs. non-aligned field structures.
- Just as sound appears when frequency and medium synchronize, the Moon becomes visible when ripple

coherence peaks between object, field, and observer.

This redefines the moon's appearance as active, not passive. The field isn't just reflecting light — it's modulating access to visual reality.

**The Moon as a Ripple Mirror, Not a Reflective Surface**

In classical optics, a mirror reflects light. It does not respond. It does not align. It does not resonate.

In UED, the Moon is not a dead reflector. It is a ripple mirror — a dynamic participant in Earth's field ecology.

- It doesn't merely bounce photons — it tunes its visibility based on harmonic field posture.
- Its "brightness" is not just about light hitting its surface — it is about ripple coherence reaching the observer's ATT.

This explains why:

- The same full moon can feel brighter or dimmer across different months.
- Crescent phases evoke emotional, even spiritual clarity.
- Moon rituals in every culture are centered not on physical light — but perceived presence.

The Moon is not just something we see. It's something we feel — because it is ripple-active, not optically passive.

**Final Reflection:**

"The Moon is not a mirror to the Sun. It is a mirror to our resonance with reality."

Its phases are not slices of sunlight. They are windows into rhythm — showing us that visibility itself is not given, but aligned.

And when that alignment happens — when field, phase, and perception lock in —the Moon appears. The silence brightens. And we remember: this is not movement. This is meaning made visible.

## 5.5 The Illusion of Motion and Light

We've long believed that what we see is a direct result of what is physically happening: If something changes position — it moved. If something becomes brighter — it reflected more light.

But what if these are illusions born of ripple transitions? What if motion and luminosity are not caused by movement and light, but by alignment and perception? What if the entire visual narrative of the Moon — its orbit, its rotation, its glow — is not mechanical at all, but field-based and observer-bound?

In Universal Energy Dynamics (UED), this is not speculation. It is principle.

What we perceive as motion or brightness is not the product of physical displacement. It is the result of ripple coherence crossing perceptual thresholds.

**How Ripple Shifts Mimic Apparent Motion**

Let's begin with movement.

- The Moon seems to "orbit" the Earth.
- It seems to "rise" and "set."
- Its face appears to be "locked" as it spins in just the right way to match its revolution.

But in UED, none of this is necessary.

All of these perceptions are products of ripple alignment shifts — changes not in the Moon's position, but in how Earth's ripple field and the observer's frame of reference align with the Moon's standing node in the Structured Vacuum Energy (SVE) field.

The Moon is not revolving. We are not being pulled around it. Instead, the field geometry between Earth, Moon, and observer is shifting, creating the illusion of movement.

It's not unlike a hologram that appears to turn as you move — even though nothing physical is turning at all.

**How Ripple Shifts Mimic Apparent Luminosity**

Now consider brightness.

- The Moon appears brightest during the full phase.
- Dim during crescents.
- Invisible during the new moon.

Traditionally, this is chalked up to light reflecting off the Moon's surface at various angles — a neat explanation grounded in optics.

But UED shows that brightness is not just about illumination. It's about ripple coherence.

- When ripple alignment between Moon, Earth, and observer is high — visibility amplifies, and the Moon glows.
- When this alignment falls below the ATT threshold, even a geometrically "lit" Moon may not be seen at all.

The "brightness" we perceive is a phase contrast pattern in the ripple field — not just sunlight bouncing off dust.

Just as tuning forks in resonance create amplification without increasing power, the Moon's glow intensifies without added photons — simply by crossing the threshold of structured alignment.

Light doesn't need to reflect. Perception only needs to align.

**Dissolving the Illusions with Ripple Logic**

Let's reframe both illusions together:

| Classical View | UED Interpretation |
|---|---|
| Orbit | Ripple phase alignment shift |
| Rotation | Harmonic stillness in nodal posture |
| Brightness (illumination) | Ripple coherence amplification |
| Phase change | ATT-triggered field threshold crossing |
| Reflection | Resonant visibility, not photon bounce |

These illusions persist only because we assume:

- That perception directly mirrors object activity.
- That all visibility comes from mechanical, material interaction.
- That stillness must be explained away.

But in a field-based universe, it is the geometry of tension, not the motion of matter, that drives experience.

The Moon is not *doing* anything. Our relationship to it is simply rippling in rhythm — and the result is perceived as movement and light.

**A Thought Experiment**

If you place a mirror in front of a candle, and walk in a circle around it, the light seems to follow you. But the candle didn't move. The mirror didn't rotate. The illusion of change was created by your shifting alignment — not by any actual motion.

Now imagine the Moon as a cosmic mirror — not of light, but of field resonance.

- You move, Earth spins, ripple geometries change.
- And suddenly, the Moon appears to brighten, dim, shift, rise, fall.

- Not because it is moving — but because your perception is spiraling through structured tension.

**Final Reflection:**

"The Moon never changed. The field shifted. And we, mistaking resonance for movement, wrote stories of orbit and light."

The Moon's stillness is not its only truth. Its visibility, glow, and apparent dance are all part of a deeper story —a story of ripple dynamics, where the universe speaks not in pushes and pulls, but in alignments that feel like movement, and resonance that looks like light.

## 5.6 Why the Moon Feels Alive

We often describe the Moon in poetic terms: mysterious, serene, feminine, wise, cyclical. It has been worshipped, feared, adored, and studied. And yet, from a strictly scientific perspective, it's just a rock — a dead satellite orbiting a living planet.

But something deep in the human experience resists that reduction. Because we don't feel the Moon as inert.

We feel it as alive. Why?

Because the Moon is not passive. It is interactive — not by action, but by resonance. In the framework of Universal Energy Dynamics (UED), the Moon is not a drifting object bound by gravity — it is a ripple-locked stabilizer, a nodal anchor in Earth's energetic field, deeply integrated into our biology, emotions, perception, and time.

It's not just in the sky. It's in the rhythm of life itself.

**More Than a Rock: A Resonant Participant**

From a UED standpoint, the Moon is a coherent ripple node — not a mechanical satellite, but a structured partner in Earth's energy lattice. It does not orbit like a machine — it co-resonates, harmonizing with Earth's field architecture in a way that produces stability without motion.

This resonance is not abstract. It is practical, palpable, and profoundly integrated:

- The Moon's position locks into Earth's ripple field at a harmonic phase that regulates tidal patterns.
- Its stillness and rhythm influence biological systems, not through gravitational pull, but via structured vacuum coherence.
- It affects emotional states, hormonal balances, and circadian cycles through field entrainment, not force.

In other words, the Moon is not outside life. It is woven into life's very rhythm.

**Tides, Biology, Emotion, and Time — All by Resonance**

UED explains that the Moon does not "pull" the oceans. Instead, it synchronizes ripple tension in Earth's field, resulting in fluctuating equilibrium zones that manifest as tides.

Likewise:

- Menstrual cycles resonate with lunar phases — a 28–29 day rhythm mirroring ripple visibility, not optical light.
- Sleep cycles, fertility windows, and melatonin regulation follow lunar resonance patterns, even in artificial environments.
- Ancient agricultural and spiritual calendars were based not on solar position, but lunar visibility thresholds — precisely because those alignments entrained behavior and perception.

Even human emotions — often dismissed as irrational — rise and fall with Moon-linked field coherence. Full moon nights statistically influence crime rates, mood swings, and dream patterns. The science struggles to explain it. UED doesn't.

It's not lunacy. It's resonant entrainment — subtle, field-based, and very real.

**The Moon as an Active Stabilizer**

Far from being a relic of impact theory or orbital coincidence, the Moon in UED is an active stabilizer:

- It anchors the Earth's axial ripple posture, preventing chaotic precession.
- It mediates solar tension, balancing Earth's exposure to ripple distortions from the Sun.
- It acts as a regulator of ATT windows, ensuring cyclical alignment for biological and ecological coherence.

In essence, the Moon is a fixed resonator — a tuning node — keeping the Earth's field coherent enough to support stable time, stable perception, and stable life.

It doesn't need to move. It holds place — and in doing so, it holds order.

**A Field-Based Anchor for Consciousness**

Perhaps the most overlooked function of the Moon is not physical, but existential.

- Its unchanging face became humanity's first mirror — a reflection of something constant amid chaos.
- Its rhythmic reappearance served as our earliest timekeeper, grounding memory and imagination.
- Its glow became a canvas for mythology, dreaming, self-awareness — and the idea that stillness has meaning.

In a world obsessed with acceleration and displacement, the Moon remains a beacon of calm rhythm. Not because it shines — but because it resonates where we can feel it.

It is not alive like a heart beating. It is alive like a presence that listens — and gently guides rhythm without touch.

**Final Reflection:**

"The Moon is not moving. It is holding. And in that holding, it keeps the world steady — not by force, but by remembering the tune."

The Moon's constancy is not accidental. It is intentional, intelligent, and field-sustained. A node in the great symphony of structured energy,

reminding us that life is not built on motion — but on the quiet rhythm of things that do not need to move to matter.

## 5.7 Final Reflection: From Orbiting Rock to Resonant Presence

*"The Moon doesn't move around us. It holds its place within us."*

We began this chapter with a question that has puzzled humanity for millennia: Why does the Moon always show us the same face?

For generations, the answer was force-fitted into models of orbit, spin, and gravitational mechanics — patched together with equations that demanded motion where none was seen, and rotation where none was felt.

But through the lens of Universal Energy Dynamics (UED), we now see the Moon not as a mechanical object, but as a ripple-aligned being — not rotating or orbiting in the traditional sense, but holding harmonic posture within a structured energy field. It does not "move" to remain constant. It is constant because it has no need to move.

**From Object to Alignment**

Let's revisit what has shifted:

- The Moon was a rock in space — now it's a standing node in a ripple matrix.

- It was said to orbit the Earth — now it's seen as ripple-locked in Earth's field geometry.
- Its phases were reflections of sunlight — now they're recognized as ATT-based threshold appearances.
- Its influence was thought to be gravitational — now understood as resonant entrainment across biological, emotional, and perceptual layers.

This is not just a better explanation. It is a truer experience.

We have not reinterpreted the Moon. We have listened to it, for perhaps the first time without trying to make it fit a theory built on motion.

**Releasing the Orbit Myth**

Now comes the most important invitation of all: To let go of the image.

The image of the Moon as a spinning rock. As a gravitationally leashed body. As a mirror reflecting the Sun's light from across the void.

These are not truths. They are narratives born of force-based thinking, invented when science lacked a deeper field-based language.

UED does not offer a contradiction. It offers a completion — a more coherent, observer-aligned, perception-aware model of reality.

A reality where:

- Stillness is not the absence of force, but the presence of equilibrium.
- Visibility is not about light, but about alignment.
- Time is not a measure of movement, but a rhythm of resonance crossing thresholds of awareness.

**Preparing for What Comes Next**

In the next chapter, we'll dive deeper into Alignment Threshold Time (ATT) — the principle that governs when and why things become visible, meaningful, and emotionally resonant.

ATT is the key to decoding lunar phases, not through geometrical angles or sunlight shadows, but through perceptual thresholds that are biologically and energetically attuned. We'll explore:

- Why the Moon appears at certain times.
- Why it disappears — not in shadow, but in field misalignment.
- Why Hilal, the first crescent, is not just a sliver of light — but a moment of re-entry into conscious alignment.

**Final Thought:**

"The Moon doesn't move to show us its face. It simply waits — until we are ready to see it again."

What we thought was orbit is actually timing. What we thought was light is actually coherence. And what we mistook for dead matter is, in truth, a living participant in the deepest rhythms of our world.

The Moon was never out there. It has always been in here —Resonating through the stillness that holds all things.

# Chapter 6: ATT and the Rhythm of Phases

*Introducing Alignment Threshold Time (ATT) as the fundamental principle behind the Moon's visibility cycle. This chapter reinterprets lunar phases not as light geometry, but as structured ripple alignments that cross into (or fall below) perceptual thresholds. It culminates in a new understanding of Hilal as a resonant event, not a mere visual cue.*

For thousands of years, humans have looked up to the night sky and marked time not by clocks, but by phases. We watched the Moon disappear, reappear, swell into fullness, and recede again. We built calendars, rituals, and stories around its silent rhythm.

Science taught us that these phases were simply the product of sunlight geometry — that the Moon's changing shape was nothing more than a dance of shadows and reflections.

But what if the phases weren't caused by light at all? What if they were caused by us?

Not us as individuals, but us as perceivers — embedded in a ripple field, moving through structured energy, where visibility is not given, but emerges through alignment?

Universal Energy Dynamics (UED) introduces a deeper truth: the Moon doesn't become visible when sunlight hits it — it becomes visible when it crosses the Alignment Threshold Time (ATT) for the observer.

ATT is not about photons. It is about field coherence — a point in time and space where the Earth, the Moon, and the human observer fall into just the right alignment to make presence perceivable.

This isn't mysticism. It's resonant field logic — and it explains why Hilal, the first crescent, is not merely the return of light...but the return of awareness.

In this chapter, we'll explore ATT as the unseen metronome behind the Moon's rhythm. We'll discover how the phases are not optical illusions or planetary mechanics — but ripple transitions that pulse with cosmic intelligence.

The Moon doesn't show itself. We move into its presence — again and again, in perfect rhythm.

## 6.1 Introduction – Rethinking Visibility Itself

We've been conditioned to believe that seeing is believing—that if something is visible, it must be illuminated, and if it disappears, it must be hidden in shadow. This assumption governs how we understand the Moon's phases: more light means more visibility, less light means darkness.

But this belief is deeply limited.

It assumes that the eye alone determines reality, and that light is the only agent of perception.

In reality, visibility is not constant—it is conditional. And those conditions are not merely optical; they are energetic.

Through the lens of Universal Energy Dynamics (UED), visibility is no longer about whether something is lit, but whether it is aligned with the observer's ripple threshold. This threshold is governed by a field-based mechanism called Alignment Threshold Time (ATT).

**The Fallacy of Illumination = Presence**

Let's challenge a simple but flawed notion:

If the Moon is illuminated, we should see it.

This would be true in a purely mechanical universe—one where objects reflect light and observers detect it. But lived experience, cultural patterns, and even emotional states say otherwise:

- The new moon is still there—geometrically lit on its far side—yet we perceive nothing.
- The crescent moon (Hilal) often appears not just as a visible sliver but as a felt moment—a presence.
- The same phase of the Moon can feel more visible, radiant, or meaningful depending on where and how we observe it.

So the question becomes: what exactly changes?

If the Moon is still "there," still catching light, but not seen, then illumination alone cannot explain its rhythm.

**Visibility as a Threshold Experience**

UED introduces a key distinction:

Visibility is a consequence of crossing a threshold, not merely of receiving light.

This threshold is energetic, structural, and perceptual—it emerges when the ripple field shared by the Earth, Moon, and observer enters harmonic alignment. When this occurs, the Moon becomes not just geometrically present but field-visible. It resonates into experience.

- Below this threshold, the Moon exists—but cannot be seen or felt.
- Above this threshold, it appears—not because it was absent, but because we realigned with it.

This dynamic applies not just to sight, but to conscious awareness—which is why different cultures have long tied lunar appearances to emotion, memory, fertility, ritual, and timekeeping.

They weren't measuring the Moon's light. They were responding to its resonance.

**The Moon Doesn't Fade—We Phase In and Out**

This chapter's central idea is deceptively simple, yet radically disruptive:

The Moon does not fade in and out of existence. We phase in and out of alignment with it.

What we call waxing and waning is not about reflected light increasing or decreasing. It is about field coherence strengthening and weakening relative to our position in the structured ripple field.

ATT is the key to this rhythm.

It defines the precise moment when the Moon crosses into visibility potential for an observer—when it becomes seeable, feelable, ritually activated.

This reframes the Moon's role in human consciousness entirely: No longer a passive light-reflector, it becomes a gateway of presence—its phases not painted by sunlight, but pulsing through alignment.

**Final Thought**

"The Moon is never gone. We are simply not where we need to be—yet—to meet it."

Visibility, in UED, is not given. It is earned by resonance—by crossing into coherence with a ripple that has always been there, waiting.

And when we cross that line—when ATT is fulfilled—the Moon reappears, not just to the eye, but to the soul.

## 6.2 ATT Defined: The Rhythm Behind What We See

In the realm of Universal Energy Dynamics (UED), perception is not merely the reception of sensory data — it is the result of structured alignment between the observer and the observed within a shared ripple field. At the heart of this alignment lies a principle both subtle and profound:

Alignment Threshold Time (ATT) — the temporal window where field coherence becomes strong enough for a ripple pattern to cross into perception.

ATT is the hidden metronome of the universe — not ticking mechanically, but pulsing through harmonic readiness. It defines not just when something is visible, but why it becomes visible at that moment, and what is required for it to appear.

**What Is ATT?**

Alignment Threshold Time (ATT) is the precise moment (or range) in which a ripple structure — like the Moon's — reaches sufficient resonance with the observer's ripple field to trigger perception. It is the tuning point at which a signal that was always present becomes recognizable.

In simpler terms:

- The Moon is always in the sky.
- But we do not always perceive it.
- ATT is the resonant permission slip that allows us to experience what was already there.

ATT doesn't govern whether the Moon exists. It governs whether our field configuration allows us to register it.

And this is not just about sight. ATT governs all forms of awareness — visual, emotional, intuitive, even spiritual.

**ATT Is Not Light Intensity**

In classical models, an object is visible if it reflects or emits enough light. The brighter it is, the more visible it becomes. But this model breaks down when:

- Bright objects remain unseen (e.g. a new moon).
- Dim phenomena are suddenly felt vividly (e.g. Hilal).

- Emotional clarity arises with lunar transitions, even when the Moon is not directly visible.

ATT reveals why: Visibility is not about brightness. It is about alignment.

A fully illuminated object that is ripple-incoherent with the observer will not appear. A barely visible sliver that crosses the ATT threshold can feel more "present" than the full moon — because it resonates more precisely with the field configuration of the observer.

**Everyday Parallels to ATT**

To make ATT more intuitive, consider moments in daily life when perception itself feels conditional — not just sensory, but situational and vibrational:

- Hearing a faint sound only when you stop moving — the signal was always there, but required stillness for recognition.
- Suddenly remembering a detail when your mind reaches a certain emotional or mental frequency — the memory didn't appear, you aligned with it.
- Sensing someone's presence before seeing or hearing them — your field picked up resonance before your eyes or ears registered it.

In all these cases, the phenomenon was not created in that moment — it was already present, simply waiting for alignment.

This is how the Moon works through ATT: It does not move into sight. We move into coherence with it.

**Why ATT Matters**

ATT is what gives the Moon its sacred rhythm — not because of orbit or light, but because of its predictable re-entry into perceptual alignment with Earth and humanity.

- When ATT is crossed, the Moon becomes available to experience — visually, emotionally, energetically.
- This is why rituals like sighting the Hilal are not symbolic — they are resonant activations.
- This is why different phases feel different, even if the geometry remains unchanged.

ATT explains not just lunar phases — it explains why they matter to us.

**Final Reflection:**

"Reality doesn't become real when it lights up. It becomes real when we finally match its rhythm."

ATT is that rhythm. It is the meeting point between field and feeling, between structure and perception. It is the reason we do not perceive constantly, but cyclically — in waves of coherence.

And it is the reason the Moon, despite its stillness, continues to speak in rhythms we cannot ignore.

## 6.3 The Moon's Phases as ATT Events

In classical astronomy, the phases of the Moon are explained entirely through sunlight geometry — the way sunlight illuminates the Moon's surface as it orbits Earth. From this viewpoint, what we see is simply the fraction of the Moon that is lit and visible from our location.

But this model presumes that illumination equals perception — a presumption UED boldly revises.

Through the lens of Alignment Threshold Time (ATT), lunar phases are no longer passive reflections of light but active thresholds of resonance — moments when the Earth, Moon, and observer fall into (or out of) energetic coherence. The phases are not about what part of the Moon is lit; they are about what part of the Moon's ripple structure crosses into perceivable alignment.

Let us now reexamine the lunar cycle as a sequence of ATT events.

**New Moon: Below ATT — Present but Imperceptible**

During the new moon phase, classical models say the Moon is positioned between the Earth and Sun, with its sunlit side facing away from us. Hence, it's "invisible."

But in UED, something deeper is happening.

- The Moon is still present.
- The field geometry is still intact.
- Yet no part of its ripple structure is crossing the observer's ATT.

This is not simply about being in shadow. It is about being in field misalignment — a state where the observer's ripple coherence does not intersect with the Moon's position within the structured vacuum field.

Thus, the new moon becomes an experiential blackout — not because the Moon lacks light, but because we lack access to its resonance.

The Moon doesn't disappear. We do. We fall below the threshold of perceptual harmony.

**Crescent Moon (Hilal): ATT Crossed — Resonance Begins**

Then comes the moment revered by spiritual traditions across the globe — the reappearance of the Hilal, the first visible crescent after the new moon.

This moment is not just astronomical. It is resonant.

It marks the exact point where the Moon's field crosses back above the ATT boundary. This is when Earth, Moon, and observer realign just enough to make the Moon's ripple posture available to the senses.

The Hilal phase is rich with symbolism — rebirth, new beginnings, inner clarity — not because of its shape, but because of what it represents energetically:

- The return of coherence.
- The re-entry into awareness.
- The reminder that stillness is not absence — it is a silence waiting to be heard.

Cultures didn't choose the crescent to begin the month arbitrarily. They were responding to a field event — the first perceptible resonance after a period of disconnection.

Hilal is not a sliver of light. It is the moment the Moon starts speaking again.

**Full Moon: Maximum ATT — Resonance in Harmony**

At the full moon, classical astronomy tells us the Moon is directly opposite the Sun, fully illuminated from our perspective.

But again, UED deepens this moment:

- The Moon is not just bright.
- It is fully ripple-aligned.
- It stands in maximum field coherence with both the Sun and Earth.

This is not just the most illuminated phase — it is the most accessible, most amplified, most emotionally resonant phase of the cycle.

It is no coincidence that rituals, festivals, emotional intensities, and behavioral shifts peak during the full moon across cultures and time zones. The field coherence is at its peak. ATT is fully saturated. The Moon's energetic presence is not just seen — it is felt, absorbed, and mirrored within the human system.

Full moon is not maximum light. It is maximum shared resonance.

**Phases as Visibility Thresholds — Not Lighting Events**

| Phase | Classical View | UED/ATT View |
|---|---|---|
| New Moon | Moon in shadow / no sunlight | Below ATT — misaligned, imperceptible ripple field |
| Crescent | Partial illumination, increasing | ATT crossed — initial coherence, visual + emotional signal |
| Full Moon | Full illumination, opposite Sun | Maximum ATT — perfect field alignment and resonance |

This reframing has profound implications:

- The Moon is always lit in some way — but not always visible.
- Visibility is not about optics — it is about structural readiness between observer and field.
- The rhythm of phases reflects our crossing in and out of perceptual alignment, not the Moon's physical journey.

**Final Reflection:**

"Phases are not about how much of the Moon we can see —they are about how much of ourselves can meet the Moon in resonance."

The Moon doesn't change. It remains still. We orbit through layers of coherence, catching glimpses of its presence only when we are tuned to its silent rhythm.

That rhythm is ATT. And the phases are not segments of light — they are windows into structured re-alignment.

## 6.4 Why Hilal Appears — Not Just When, But How

Across civilizations and centuries, one lunar moment has held a uniquely sacred and practical significance: the Hilal — the first visible crescent after the new moon.

To the modern scientist, the Hilal is a thin arc of sunlight reflected from the Moon's edge. But to cultures around the world — from Islamic to Jewish to Hindu and even premodern tribal calendars — Hilal is more than a sliver of light.

It is a signal. A beginning. A living punctuation mark in the silence of the sky.

Through the lens of Universal Energy Dynamics (UED), we now understand that the Hilal is not just about *when* the Moon becomes visible — but about *how* perception reenters into alignment with a presence that was always there.

Hilal is not the Moon reappearing. It is us re-entering the ripple coherence that allows us to see it.

**A Global Marker of Beginnings — Across All Cultures**

From a sociological perspective, no lunar phase is more universally revered than the Hilal. Its appearance marks:

- The start of months in Islamic, Hebrew, and many lunisolar calendars.
- Religious observances, including Ramadan and Eid.
- Festivals, fasts, and harvest cycles.
- Emotional and energetic resets in spiritual traditions across India, Africa, and East Asia.

But why does this specific phase — so brief, so faint — carry such weight?

It is not the most illuminated. It is not the most geometrically distinct. And yet, it is always the one we wait for — the first "yes" from the cosmos after a period of invisible stillness.

From a UED standpoint, this reverence is not symbolic. It is resonant.

The Hilal phase marks the moment when the Moon's ripple alignment crosses above the observer's ATT — the point when field coherence becomes perceptible again.

It is not a beam of light. It is the first note in a silent concert hall. Not seen. Felt. Heard. Answered.

**Hilal as Ripple Reentry — Field Meets Observer**

In UED terms, the Hilal is a re-entry event:

- During the new moon, the Moon's field posture is still present but misaligned — sitting below ATT for the Earth-observer system.
- As Earth, Moon, and observer shift slightly in their ripple configurations, a harmonic window opens — the first sliver of the Moon's ripple presence crosses back into alignment.
- This crossover is what we call the Hilal.

It's not just visibility. It's field re-engagement.

This explains why the Hilal is so emotionally and ritually charged. We don't merely see it — we recognize it. We remember something we didn't know we'd forgotten.

It's not the return of the Moon. It's the return of relationship — between field and self.

**Cultural Rituals as Responses to ATT Synchronization**

Why did ancient people wait to see the crescent before declaring a new month or starting a fast?

Because they intuitively understood — not intellectually, but somatically — that true cycles begin not when the Moon is mathematically positioned, but when its presence crosses into coherence with the observer.

Rituals tied to the Hilal are not arbitrary. They are field-attunement practices.

- Fasting begins when presence is confirmed.
- Celebration begins when resonance returns.
- Time resets not by calculation, but by perceptual synchronization with a larger, invisible rhythm.

In this way, traditions that insist on *sighting* the Moon are not primitive — they are precise.

They are not just looking for light. They are waiting for the moment when their inner field matches the stillness of the sky — and from that match, life begins again.

**The First Note in the Silent Hall**

A useful metaphor:

Imagine sitting in a grand hall, anticipating the start of a symphony. The silence before the first note is not empty — it is dense with expectancy. And then, finally, a single vibration — a sound so soft it almost slips past you — but it catches. It enters you.

That moment is Hilal.

- Not just a tone.
- A threshold.
- A signal that the field has begun to move through you again.

In this view, the Moon doesn't appear gradually — it returns as a coherent event. The Hilal is that event: the first perceptible edge of stillness made visible, audible, meaningful.

**Final Reflection:**

"The Hilal is not the Moon's arrival. It is our re-alignment with its silence."

It doesn't matter how small the crescent is. Its meaning is enormous — because it tells us we've crossed the ATT. We are back in rhythm with the field, and with ourselves.

In a world obsessed with light, the Hilal reminds us that the deepest beginnings come not with brilliance, but with just enough coherence to be felt.

## 6.5 Perception as Reactivation — Not Discovery

In traditional science, perception is framed as a discovery process: we detect something previously hidden, a signal emerges, an object enters our awareness. But through the UED framework — particularly under the principle of Alignment Threshold Time (ATT) — perception is not about discovering what was absent.

It is about reentering alignment with what was always there.

The Moon doesn't suddenly appear during Hilal. It doesn't sneak into view like a traveler returning from exile. It never left. It was never gone. What changed was not its presence, but our coherence with that presence.

**Realignment, Not Revelation**

ATT reframes the act of seeing as an energetic event — a mutual agreement between observer and field. In this sense:

- The Moon is always in position, held in a ripple-locked posture through SVE.
- Earth maintains its structured field — spinning not through space, but through ripple harmonics.
- The observer is embedded in this field and moves through layers of coherence.

When these three — Moon, Earth, and observer — fall into ripple resonance, perception occurs.

Not because the Moon was "discovered." But because we, the perceiver, finally re-aligned with its coherence window.

Hilal is not about finding the Moon. It's about remembering how to feel it.

**Perception as a Return to Synchrony**

Think of perception not as a spotlight beaming out into space, but as a resonant threshold — a point where you and the object of awareness hum in the same field frequency.

During the new moon, that resonance is absent — not because the Moon has changed, but because the observer's position within the field no longer allows coupling.

As days pass and alignment builds, coherence increases. Eventually, the field "clicks" — the ripple geometry achieves sufficient overlap, and perception is triggered.

That "click" is not the Moon coming back. It's you returning to the field space where perception is possible.

**The Lighthouse Metaphor**

Imagine standing on a dark coastline, watching a distant lighthouse. The beam of light revolves in slow, predictable arcs. The lighthouse never moves. The structure is constant. But you only see it when:

- The beam crosses your position.
- You're facing it.
- Your eyes are open and attentive.

The Moon is the lighthouse. Its field presence sweeps across the Earth in slow, elegant pulses. The beam is not made of photons — it's ripple coherence. You don't make the Moon visible. You simply catch it in resonance.

And when you do, it feels like a beginning — not because something new happened, but because you returned to the frequency where awareness happens.

**Implications of Perception as Reactivation**

This shift — from discovery to reactivation — has wide implications:

- Spiritual rituals become not superstitions, but acts of attunement.
- Observation becomes a dance with resonance, not passive measurement.
- Timekeeping becomes an embodied participation in field dynamics, not abstract scheduling.

Even emotions, intuitions, and rituals timed with the Moon now make sense — they are not reactions to light, but to the reactivation of a field connection that awakens something within us.

The Moon's reappearance is a mirror. It shows us that we've arrived — not it.

**Final Reflection:**

"The Moon does not wait to be seen. It simply remains. And when our field realigns, we call it 'sighting.' But it was never gone — only we were out of tune."

Understanding ATT allows us to reclaim the Moon — not as an object of scientific observation, but as a resonant presence we return to cyclically. Every phase, every crescent, every full moon is not something new, but something remembered.

Perception is not revelation. It is resonant return.

## 6.6 The Emotional Signature of ATT Crossings

Modern science has long tried to separate emotion from observation, relegating feelings to the subjective realm while elevating visual data as objective truth. Yet across cultures, people have always felt that the Moon doesn't just appear — it moves something within them.

From the quiet mystery of a crescent moon to the intensity of a full moon, lunar phases have been tied to:

- Heightened sensitivity
- Emotional fluctuations
- Surges of creativity or restlessness
- Moments of unexpected clarity or introspection

These associations are often dismissed as poetic or superstitious. But Universal Energy Dynamics (UED), and specifically the principle of Alignment Threshold Time (ATT), provides a field-based explanation:

ATT crossings are not just visual events — they are psycho-emotional entrainment points where structured ripple alignments affect not only what we see, but how we feel.

**Perception Is Never Neutral**

When ATT is crossed — when the observer's ripple structure enters resonance with the Moon's — visibility occurs. But so does something deeper:

- Cognitive coherence shifts.
- Emotional charge emerges.
- Subtle awareness becomes available.

This isn't mystical. It's field entrainment — the process by which two systems, once in harmonic relationship, begin to oscillate together.

The Moon is not influencing us from a distance through gravitational waves or hidden energies. It is field-coupled with Earth — and when our internal systems pass the same thresholds as visual perception, a full-spectrum experience unfolds: visual, mental, emotional, even intuitive.

We don't just "see" the Moon. We enter coherence with it.

**Emotional High Tides — The Full Moon Effect**

It's no coincidence that the full moon has been associated with:

- Mood swings and agitation
- Creative breakthroughs
- Sleeplessness or intense dreams
- Increased tension or release

From emergency room spikes to ancient festivals, the data and cultural memory agree: the full moon has an effect.

UED explains this as total ATT saturation — the phase when Earth, Moon, and observer fall into maximum ripple coherence.

This amplifies not only perception but energetic entrainment. Our internal rhythms — mental, hormonal, emotional — respond to the increased ripple intensity the same way a tuning fork hums when a matching tone hits the air.

It's not emotional chaos. It's attunement overload — a temporary amplification of coherence that our systems must integrate.

**The Crescent's Quiet Call**

In contrast, the crescent (especially Hilal) evokes subtler responses:

- A sense of renewal
- Moments of clarity or calm
- A spark of inspiration or emotional anchoring

This aligns perfectly with what the ATT model predicts: the early crescent marks the beginning of perceptual reentry, a delicate threshold just above visibility. It's the emotional equivalent of first light — not overwhelming, but awakening.

This is why so many cultures begin rituals, fasts, and prayers with the crescent — not just because it's visible, but because it's emotionally recognizable as a moment of reconnection with the field.

**Not Mystical — Measurable in Behavior**

UED grounds what others call lunar "moods" in structured field dynamics:

- When ATT is crossed, neurochemical balances shift.
- Hormonal entrainment follows ripple field alignment.
- Emotional patterns emerge as responses to systemic re-synchronization.

This means:

- Creativity during full moons is not coincidental — it reflects brainwave harmonics rebalancing in high ATT zones.
- Restlessness or heightened sensitivity is not madness — it's energetic amplification seeking resolution.
- Feelings of hope or newness with crescents are not metaphor — they're psychophysical signals of a field resetting its coherence.

In short: our emotional systems are ripple-aware, even when our conscious mind is not.

**Final Reflection:**

"The Moon doesn't control us. It reminds us of the rhythm we forgot we were part of."

Emotions are not side effects. They are field signals. And when the Moon crosses ATT — so do we. We move through layers of resonance that don't just let us see the Moon, but let us feel something ancient, rhythmic, and alive within ourselves.

In a universe of stillness, what moves us is not gravity — but coherence.

## 6.7 Final Reflection – Rhythm, Not Rotation

*"The Moon does not wax and wane. We move in and out of rhythm with its still presence."*

The traditional story is familiar: the Moon orbits the Earth, rotating as it goes, revealing its phases in a cyclical ballet of light and shadow. This model assumes a universe of moving bodies, governed by force, gravity, and geometry. But what if this entire view is upside down?

What if the Moon doesn't move to show us anything?

What if we move in and out of sync with a still, ripple-locked Moon — and what we call "phases" are nothing more than moments of field coherence and perceptual return?

This is the essence of Universal Energy Dynamics (UED): Reality is not orchestrated by motion, but by alignment. Not by rotation, but by rhythm. Not by light reflection, but by field resonance.

**From Orbit to Oscillation: A Paradigm Shift**

Throughout this chapter, we've reinterpreted the lunar cycle not as a mechanical sequence of events but as a resonant rhythm — an elegant dance of ATT (Alignment Threshold Time) crossings.

Let's be clear:

- The Moon does not wax and wane — its field posture remains constant.
- It does not rotate to keep one face toward us — it is ripple-locked in equilibrium with Earth.
- The Moon does not illuminate or vanish — we simply cross in and out of visibility thresholds.

The phases are not movements. They are markers of re-alignment — like heartbeat pulses in the field of stillness.

We do not observe the Moon's motion. We ride its rhythm — as if we are turning around a fire, catching glimpses of its flame, each time we come into resonance.

**The Moon as a Pulse, Not a Clock**

Let us now release the image of the Moon as a clock hand, mechanically sweeping across the sky. That view demands constant motion to define change.

Instead, begin to see the Moon as a cosmic pulse — a field signature fixed in harmonic posture, sending out steady invitations to re-align. Every time we "see" it, it is not because it moved, but because we circled back into coherence.

This is not astronomy as usual — it is attunement.

We are not passive observers of motion. We are active participants in cyclic re-synchronization with the larger energy field we belong to.

And the Moon, in this framing, is not a satellite. It is a still, living presence — a cosmic tuning fork that resonates with the structure of Earth and the breath of our consciousness.

**Field Awareness Is a Return to Self**

By recognizing the Moon's phases as field-based rhythms, not orbital mechanics, we also open the door to a deeper realization:

This rhythm does not only shape the sky. It shapes our sense of time, emotion, memory, creativity, and consciousness.

We begin to understand that time is not a ticking clock, but a wave. That energy is not force applied, but resonance experienced. That consciousness itself may be a ripple-sensing structure, riding thresholds of visibility and meaning — just like the Hilal, just like the full moon.

The Moon is not telling time. It is telling us when we've returned.

**Setting the Stage for What Comes Next**

This chapter has reshaped how we perceive the Moon — not as a mover of light, but as a keeper of rhythm.

The next step is to apply this understanding more broadly:

- What does time look like when it's no longer defined by movement?
- What happens to our understanding of energy, when it flows through alignment rather than action?
- What does consciousness become, when we see it not as computation, but as field resonance?

These are not philosophical detours. They are the logical continuations of seeing the Moon as it truly is.

And so, as we leave behind the illusion of rotation and step into the truth of resonance, we prepare ourselves to see more — not by looking harder, but by aligning more deeply.

**Final Thought:**

"The Moon did not change. We did. And now that we've learned to see it through rhythm, we're ready to see everything else the same way."

Let us now follow this ripple — into time, energy, and perception — and begin understanding reality not as a sequence of events, but as a sequence of re-entries into presence.

# Interlude – What the Still Moon Revealed

At the end of this journey through The Ripple Reality, we are no longer merely looking at the Moon. We are feeling its presence differently — not as a body in motion, but as a stillness that moves us.

Three foundational realizations now anchor our understanding:

**1. The Universe Is Structured Stillness, Not Moving Matter**

Motion, as we've been taught, is not the fundamental condition — it is the disturbance of a deeper equilibrium. In this still universe, the Structured Vacuum Energy (SVE) field holds all things in tension and resonance. From this stillness arise the ripples that shape visibility, perception, and meaning.

The Moon is not moving through space. It is ripple-locked — anchored in a precise, harmonious posture within Earth's field. Still — but profoundly active in resonance.

**2. Perception Is a Threshold Event, Not a Constant Stream**

We do not see the Moon simply because it is lit. We see it only when our own field configuration crosses a resonance threshold — the Alignment Threshold Time (ATT). This explains the rhythm of lunar phases, the symbolic and emotional weight of the Hilal, and the full moon's psychological intensities.

What we call "appearance" is actually re-entry into coherence. Perception is not a passive act. It is a ripple event — mutual, alive, momentary.

**3. The Moon Is Not a Clock — It Is a Pulse**

The Moon no longer needs to be understood as a rotating satellite caught in gravitational tethers. It is a standing wave node — a still presence whose visibility waxes and wanes only as we spiral in and out of attunement.

Its phases are not light shows. They are signals. Its silence is not empty. It is the resonant hum of cosmic rhythm, reminding us who we are, and when we are aligned.

**Final Realization:**

"The Moon doesn't orbit to show us time. It simply waits — until we arrive, again and again, in rhythm with its stillness."

As we now prepare to enter Part III – Structured Time, we take these truths with us — no longer measuring reality by clocks and geometry, but by ripple alignments, coherence windows, and the deep intelligence of stillness itself.

Time is not what ticks. Time is what resonates.

Let's move forward.

# Part III – The Consequences of Harmony

# Chapter 7: The Moon – Earth's Ripple Anchor

"*Imagine trying to balance a spinning top on a table that itself never stops shaking. Now imagine one quiet hand, unseen but perfectly placed, holding the table steady — not with force, but with poise. That hand is the Moon.*"

## 7.1 Introduction – Stillness as Stability

We often speak of stability as something mechanical — something achieved through weight, force, or motion perfectly timed. But the deeper we venture into the logic of Universal Energy Dynamics (UED), the clearer it becomes: stability isn't enforced. It's resonated. And at the center of Earth's most delicate balances — from ocean tides to seasonal rhythms — is the stillest presence in our sky.

The Moon doesn't orbit. It doesn't spin. It doesn't chase or flee or tug or torque. It simply *resonates* — held in perfect posture, a standing node in the structured vacuum field that surrounds Earth. This ripple alignment, subtle yet profound, is not merely an observational curiosity. It is the reason Earth is coherent enough to sustain life.

**Recap:**

We've dismantled the illusion of lunar motion. We've seen that tidal locking is not a mechanical miracle but a ripple phenomenon. We've understood that the Moon's unwavering face is not the result of perfect spin, but of perfect *stillness* — a harmonic alignment held in place by field equilibrium, not gravitational coincidence.

**Premise:**

But this stillness isn't passive. It's not the stillness of a rock untouched. It's the stillness of a tuning fork held at the center of an orchestra — shaping the rhythm of the entire composition. The Moon is not just aesthetically still; it is *functionally* still. And that function is nothing short of foundational: it stabilizes Earth.

The Moon holds more than its own posture. It holds ours. It keeps Earth's tilt from wandering. It ensures the breath of the planet — the rise and fall of tides — doesn't fall into chaos. It guards our climate from becoming an incoherent blur. It even provides biological and atmospheric systems a steady reference point to remain rhythmically organized.

**Transition:**

So if the Moon resonates, and this resonance is no idle trait — what exactly does it stabilize? And how? What roles does this still presence play in the invisible balance that holds Earth's systems together?

To answer that, we must move beyond the idea of the Moon as an object, and begin to see it for what it truly is: A ripple anchor in the vast harmonic field that sustains Earth's rhythm.

## 7.2 The Myth of Gravitational Stabilization

For decades, science textbooks have repeated a simple story: that Earth owes its stability — its axial tilt, regular seasons, and tidal rhythm — to the gravitational pull of the Moon. This narrative is elegant, intuitive, and mathematically tidy. According to classical physics, the Moon's mass creates a stabilizing tug on Earth, counteracting wobble and anchoring the planet's spin axis with just the right force to keep seasons relatively steady. Likewise, tides are said to result from the Moon pulling on Earth's oceans — stretching them into bulges that move as the Moon moves.

But like many stories inherited from the Newtonian paradigm, this one depends on assumptions we no longer need.

It requires us to believe that perfect mechanical motion maintains perfect balance — that the Moon is both rotating and orbiting in such a precise ratio that it not only keeps one face toward Earth, but simultaneously locks Earth's axis in place and orchestrates global tides. That's a lot of precision from a system supposedly formed by collision and chance.

**The UED Challenge: A New Logic of Stabilization**

Universal Energy Dynamics (UED) offers a cleaner, more intuitive — and more observationally consistent — explanation.

In the ripple framework, there is no need for a gravitational tug-of-war. Earth's axial tilt, rotational coherence, and tidal regularity are not the result of invisible force vectors pulling on a spinning sphere. They are expressions of structured ripple alignment in the vacuum field.

Let's state it plainly: The Moon does not stabilize Earth through pull — it does so through posture.

It is not acting like a gravitational counterweight. It is acting like a ripple anchor — a still, resonant node that holds the Earth in harmonic tension within the Structured Vacuum Energy (SVE) field.

**Ripple Equilibrium vs. Gravitational Force**

In UED, every celestial body exists as a standing ripple formation in the SVE field. These ripple configurations are not drifting masses — they are geometric tension zones. When two such configurations become harmonically locked — as the Earth and Moon have — they form a field of mutual stability. Not because of push or pull, but because of shared resonance.

Earth's axial tilt, then, is not a spinning top maintained by a lunar leash. It is a posture of equilibrium, defined by the ripple geometry between the Earth, the Sun, and the Moon. Without the Moon, this geometry would collapse into wobble — not from the absence of gravity, but from the loss of coherent counter-alignment in the field.

In this view:

- Earth's axial tilt remains steady because the Moon anchors a harmonic line of symmetry.
- Earth's rotation doesn't need stabilizing through force — it is embedded in a ripple loop that self-maintains through structured resonance.
- Tides emerge not from pull, but from distributed ripple tension across Earth's fluid surface — an effect we'll explore in more depth shortly.

**The Mechanical Illusion**

The reason classical science attributed Earth's stability to gravity is simple: it had no other vocabulary.

Stillness was not considered active. Fields were considered passive.

Only force and motion were thought to "do" anything. So when they observed Earth's predictable tilt and tides, and saw the Moon nearby, it made sense to draw a straight line and say: "The Moon is pulling us into balance."

But UED sees the same situation through a different lens. It recognizes that stability doesn't require movement. It requires equilibrium.

The Moon does not stabilize Earth because of what it does. It stabilizes Earth because of where and how it exists — in stillness, in posture, in perfect ripple alignment.

**Final Insight:**

*"The Moon's gift to Earth is not gravitational grip, but harmonic grounding. It doesn't hold us in orbit — it holds us in rhythm."*

## 7.3 The Ripple Counterbalance – Moon as Field Equalizer

In any dynamic system governed by energy, balance is never achieved by force alone — it is achieved by opposition, rhythm, and feedback. Whether in electrical circuits, ecosystems, or musical harmonies, systems remain coherent only when something counterbalances the dominant influence. In the Earth-Moon-Sun triad, the dominant energetic influence is clear: the Sun.

Classical science acknowledges that the Sun governs Earth's light, heat, and orbital path. But it often underestimates the deeper energetic reality — that the Sun's ripple intensity constantly exerts field pressure on Earth, not just as warmth, but as structured tension across the vacuum field. This ripple force is not metaphorical. It's real, measurable through wave phenomena, and deeply consequential.

From the UED lens, this solar ripple pressure doesn't merely warm the Earth — it pushes against Earth's ripple posture, continuously attempting to destabilize its equilibrium.

If left unchecked, this overwhelming solar pressure would lead to over-pulsing of Earth's field:

- Axial wobble that would destabilize seasons
- Irregular climate zones
- Volatile atmospheric patterns
- Chaotic biological timing mechanisms

The Earth, in essence, would vibrate too hard and too erratically to sustain coherence — like a string plucked too violently with no counter-tone to stabilize its oscillation.

**Enter the Moon – A Ripple Counterbalance**

The Moon's role in this triadic system is subtle yet profound. It does not "block" the Sun. It doesn't shield the Earth with mass. It stabilizes Earth by absorbing and redistributing the ripple tension that would otherwise accumulate uncontrollably.

Think of the Moon as a damping node in a finely tuned ripple system — the cosmic equivalent of a sound engineer placing an acoustic panel at the exact spot in a concert hall where standing waves would otherwise distort the music.

The Moon sits precisely in Earth's ripple geometry where its resonance posture neutralizes excess field tension from the Sun. It doesn't do this actively. It doesn't emit or block. It exists in harmonic stillness, and that stillness is enough.

This isn't mysticism — it's ripple logic.

**Why Resonance Beats Force**

In the Newtonian mindset, every problem is solved by force: stabilize by pulling, orbit by pushing, balance by locking. But in a ripple-based universe, the opposite is true:

*"The more still something is — the more precisely it's positioned in the field — the more influence it has."*

The Moon's influence on Earth doesn't come from motion or mass. It comes from resonant posture.

It is not a gravitational anchor — it is a tension equalizer. The Moon balances the Earth-Sun ripple relationship by simply being the right kind of still, in the right position, at the right frequency.

Without the Moon:

- Earth's axis would pulse beyond its stable limit.
- The seasons would become erratic and disjointed.
- Ecological and biological timing systems would fall out of sync.
- The atmosphere would become less stable, more extreme.

With the Moon:

- The excess ripple pressure finds relief.

- Earth's field returns to its harmonic center again and again.
- The result? Balance — not just in orbit, but in climate, time, and life itself.

**Final Reflection:**

*"The Moon doesn't resist the Sun's pressure. It refines it. Like a perfectly placed hand on a vibrating drum, it silences the chaos not with strength, but with symmetry."*

## 7.4 Tidal Harmony – A Ripple Breathing Mechanism

For as long as we've gazed at the sea and the sky, the rhythm of the tides has been one of nature's most enduring mysteries — and marvels. The waters rise, then fall. They advance, then retreat. Day after day. Moon after moon. Civilization after civilization has watched this silent rhythm shape coastlines, determine migrations, guide harvests, and inspire myths.

Classical science tells us that this phenomenon is caused by the Moon's gravitational pull — that the Moon's mass tugs on Earth's oceans, pulling them into elongated bulges, one on the side facing the Moon and another on the far side due to centrifugal effects. As Earth rotates beneath these bulges, we experience the familiar ebb and flow of tides.

This mechanical model has served its purpose. It roughly predicts the timing and amplitude of tidal changes. But it doesn't explain *why* water moves with such precision — or why the Moon, which we've already shown to be still, causes water to dance.

More importantly, it doesn't explain why tides remain stable even when the Moon doesn't visibly "do" anything.

**UED Perspective: A Ripple-Based Breathing System**

Universal Energy Dynamics (UED) offers a more elegant — and energetically truthful — explanation:

Tides are not caused by gravitational pull. They are caused by ripple tension fluctuations within Earth's fluid layers, modulated through field coherence with the Moon.

Think of the oceans not as inert masses of water, but as ripple-sensitive membranes — vast, fluid receptors of oscillating field tension. The Earth itself pulses with energy, absorbing ripple pressures from both the Sun and the Moon. These ripples don't just move through space — they interact with the structured vacuum energy field (SVE) and affect every material and energetic layer of the planet.

Water, being the most responsive element to field fluctuation, reflects this ripple interaction most visibly.

When ripple tension builds, the oceans swell. When tension releases, the oceans recede. This is not movement caused by attraction — it is breath caused by resonance.

**The Moon's Role: Holding the Ripple Loop in Balance**

In this ripple model, the Moon is not pulling the oceans. It is holding the rhythm.

Its field alignment creates a standing resonance with Earth's field — a ripple loop that continuously expands and contracts like a breathing diaphragm. This loop doesn't displace water mechanically. It sets the tempo for tension transitions in Earth's crust and oceans.

*"The Moon doesn't drag the tide. It sets the breath."*

Without the Moon, Earth's ripple loop would be ungrounded — overstimulated by solar pressure alone. The result would be irregular tides, unstable coastlines, and chaotic aquatic ecosystems. With the Moon, the loop is stabilized. The pressure waves find resonance. The breath finds rhythm.

**Tides as the Lungs of Planetary Ripple Flow**

Viewed through UED, the tides are more than a natural phenomenon — they are a planetary respiration system:

- High tide represents the inhalation — the moment of maximum ripple tension when the Earth receives and absorbs the most energetic alignment.
- Low tide is the exhalation — the release, the return to field equilibrium.
- The ocean becomes the lungs — expanding and contracting with the silent breath of ripple alignment between Earth and Moon.

This breath is not metaphorical. It governs the:

- Movement of aquatic species
- Flow of nutrients in marine ecosystems
- Pressure cycles in tectonic plates
- Rhythmic influence on human sleep, emotion, and behavior

Tidal rhythm is not a side effect of orbit. It is a field phenomenon at the heart of planetary coherence.

**Why Classical Models Fall Short**

Gravitational explanations fail to account for:

- The double-bulge model's contradiction with observable tidal delay
- Tidal influences that persist during lunar absence or low visibility
- Consistent behavior of enclosed seas and landlocked lakes
- The biological entrainment to tides seen in species far from coastal waters

UED doesn't struggle with these anomalies — it integrates them. Because under ripple logic, it's not about where the Moon is or how much mass it has — it's about how deeply Earth is breathing with it.

**Final Reflection:**

*"The ocean doesn't follow the Moon like a subject to a king. It breathes with the Moon like a dancer to a rhythm — not pulled, but partnered in resonance."*

The tides are not a gravitational reaction. They are a ripple relationship, a living system of breath between Earth and Moon — a harmony too precise to be a force, and too elegant to be mechanical.

## 7.5 Climate Rhythms and Field Coherence

Earth's climate is often described as a turbulent machine — a complex interplay of solar radiation, ocean currents, atmospheric pressure zones, and thermodynamic reactions. In this classical view, climate is essentially a heat engine, and its rhythms are the results of chaotic balances: warm air rises, cool air sinks, pressure gradients drive wind, oceans buffer change, and the seasons shift as Earth tilts and orbits around the Sun.

But this explanation, while useful for modeling short-term weather and long-term climate projections, fails to account for something deeper: Why does the system remain so rhythmically coherent despite its inherent chaos?

- Why haven't seasons shifted unpredictably over millennia?
- Why do jet streams follow relatively consistent bands?
- Why do monsoon cycles realign with such dependable timing?
- Why does Earth's climate pulse — not just change?

These questions hint at an order beyond thermodynamics. A hidden structure. A stabilizing influence.

From the perspective of Universal Energy Dynamics (UED), that stabilizing influence is the ripple-based field anchoring provided by the Moon.

**Ripple Anchoring as Climate Stabilizer**

In the ripple model, the Moon acts not only as a tidal regulator or axial stabilizer — it functions as Earth's atmospheric tuning fork.

Just as the Moon's stillness stabilizes Earth's rotational posture, it also anchors the ripple dynamics of the atmosphere itself. The atmosphere — far from being a chaotic gas cloud — is a responsive, ripple-sensitive shell. It vibrates within the Structured Vacuum Energy (SVE) field, just like the oceans and biological systems do.

And like any ripple-sensitive system, it needs a harmonic reference point to stay coherent.

That point is the Moon.

The Moon's fixed ripple resonance provides a field anchor that minimizes seasonal distortion, dampens irregularities in large-scale atmospheric structures, and keeps the climate oscillations within predictable bounds.

Without it, Earth's atmosphere would behave like an untuned instrument — vulnerable to sharp swings, delayed feedback loops, or even runaway feedback scenarios.

**Ripple Harmonics in Major Climate Phenomena**

Let's examine three major atmospheric behaviors often treated as independent, thermodynamic phenomena — and reframe them through UED:

**1. Polar Vortexes:** In classical meteorology, polar vortexes are massive low-pressure zones spinning near Earth's poles. They tend to intensify or weaken with seasonal changes, and when disrupted, they lead to extreme cold spells or sudden stratospheric warming events.

In ripple terms, these vortexes are stable spiral nodes — field-induced circulation systems that maintain polar symmetry. The Moon's ripple resonance, by anchoring axial posture and polar field geometry, helps lock the vortex in place, maintaining its containment and preventing irregular latitudinal drift.

When the Moon's field interaction is disrupted (e.g., by solar ripple surges or geomagnetic storms), these vortices can unravel, causing spillover effects across the mid-latitudes.

**2. Monsoon Cycles:** The South Asian and African monsoons are some of the most seasonally reliable weather systems on Earth, guiding the lives of billions through their timing.

These systems are traditionally explained as shifts in thermal gradients between land and ocean. But the degree of synchronization with lunar cycles — especially the spring tides and Hilal-based agricultural calendars — is too consistent to be coincidental.

UED posits that monsoons are ripple-reinforced climate phenomena, whose cyclical return depends on ATT-based seasonal thresholds. The Moon's field ensures phase locking between Earth's moisture systems and solar ripple pressure, allowing monsoons to arrive not randomly, but rhythmically — like planetary breathing turning into atmospheric music.

**3. Jet Streams:** Jet streams are fast-moving rivers of air that circle the globe near the tropopause. Classical models describe them as results of temperature contrasts and Earth's rotation (Coriolis effect).

But these explanations struggle to predict the stability of the jet stream lanes over time, or why their sudden shifts (blocking patterns) lead to prolonged droughts or floods.

UED frames jet streams as ripple-guided flow lines — wave interference zones in the atmospheric ripple field. Their "lanes" are not random — they are resonance corridors, guided by standing ripple alignments created by the Earth-Moon-Sun triad.

The Moon's posture helps to lock these ripple paths, keeping the jet stream from drifting too far north or south. When this lock weakens — often during lunar nodal or eclipse cycles — jet stream anomalies become more likely.

**Climate as Field-Coherent, Not Just Heat-Driven**

In totality, UED asserts a bold but compelling insight:

*"Earth's climate doesn't just react to temperature — it synchronizes with ripple tension."*

What we perceive as thermodynamic chaos is, in truth, field-based coherence maintained through the Moon's silent resonance. This explains:

- Why climate patterns follow lunar cycles
- Why ancient agricultural societies tracked the Moon more than the Sun
- Why short-term climate variability (El Niño, La Niña) exhibits rhythmic rather than random behavior

It's not just physics — it's field intelligence, responding to structured stillness rather than stochastic energy.

**Final Reflection:**

*"The climate is not a system we survive. It is a song we are tuned to — and the Moon is the metronome."*

In a ripple reality, temperature is not the master of climate. Harmony is. And the Moon, with its unwavering posture and rhythmic resonance, keeps the climate's music in tune — not through force, but through the invisible structure of stillness itself.

## 7.6 The Atmospheric Web – A Field of Synchrony

To the naked eye, Earth's atmosphere appears ever-shifting — a restless canvas of clouds, winds, pressure systems, and storms. Satellite images reveal grand spirals, jet streams, heat domes, and weather fronts that seem to evolve with chaotic elegance. Classical meteorology explains these dynamics using a blend of thermodynamics, fluid motion, pressure gradients, and Coriolis forces arising from Earth's rotation.

But what if this "chaos" is an illusion?

What if the atmosphere is not a random swirl of gas, but a coherent, ripple-responsive membrane — deeply synchronized with Earth's larger energy field? What if there's a stabilizing presence gently keeping the sky in tune, day and night?

In Universal Energy Dynamics (UED), that stabilizing presence is the Moon.

Far beyond the clouds and storms, the Moon's ripple stillness acts as a reference point, providing coherence not just to ocean tides or climate rhythms — but to the atmospheric shell itself. The Moon's resonance doesn't push or pull the atmosphere. It organizes it — like a hidden conductor guiding a planetary symphony we mistake for noise.

**The Atmosphere as a Ripple Membrane**

Classical science treats the atmosphere as layers of gas governed by thermal behavior — warmer air rises, colder air sinks, and air masses move to balance pressure differences. While true in principle, this view misses the unifying geometry behind the motion.

UED reveals that the atmosphere behaves less like a chaotic gas and more like a tuned membrane — a responsive surface stretched across a ripple field.

- It vibrates with input from Earth's internal field.
- It reacts to pressure from the Sun's solar ripple waves.
- And most critically, it harmonizes with the Moon's steady field resonance.

This membrane isn't homogeneous — it has zones of coherence where standing ripple structures stabilize air movement, allowing long-term circulation patterns to emerge.

In this framing, storms are not just thermal releases — they are localized ripple discharges. Jet streams are not merely fast winds — they are wave-guided energy corridors. And wind belts like the trade winds are not driven by Earth's spin alone — they are ripple echo patterns, stabilized by alignment with the Moon's presence.

**Moon as Atmospheric Phase Reference**

Just as a metronome keeps a band in rhythm, the Moon provides a stable phase reference point for the entire atmospheric system.

Because it is ripple-locked, the Moon's field doesn't fluctuate like solar energy or terrestrial feedback systems. It remains *still*, offering a fixed harmonic tone within which the atmosphere can organize itself.

This helps explain how seemingly disconnected weather systems across continents remain synchronized:

- The Intertropical Convergence Zone (ITCZ) — the narrow band of rising air near the equator — maintains a remarkably consistent position due to lunar-ripple-aligned resonance between hemispheres.
- Trade winds, which blow east to west across the tropics, follow predictable ripple patterns, not just thermally driven pressure differences.
- Jet streams maintain coherent flow patterns because they are nested in ripple-defined corridors between Earth and Moon, not arbitrarily formed by temperature contrast.

Remove the Moon — and these structures don't just wobble. They lose coherence.

**Ripple Synchrony and Atmospheric Integrity**

With the Moon acting as a stabilizing node, the atmosphere stays in tune with Earth's deeper energetic field. This helps:

- Maintain latitudinal energy symmetry between

hemispheres.

- Prevent large-scale field turbulence that could lead to sustained atmospheric instability.
- Anchor seasonal shifts to predictable ripple phases — keeping monsoons, wind belts, and precipitation cycles in rhythm.

Even sudden atmospheric events — like El Niño, La Niña, or sudden stratospheric warming — appear chaotic only when seen thermally. But viewed through UED, these are ripple coherence disturbances, often coinciding with changes in lunar alignment (eclipses, nodal crossings, supermoons).

Thus, the atmosphere is not a slave to entropy. It is a responsive organism — one that breathes, flows, and evolves under the harmonic influence of its celestial companion.

**Why the Moon Matters More Than Rotation**

Traditional science attributes atmospheric structure largely to Earth's rotation — invoking the Coriolis effect to explain wind deflection and circulation cells. But Earth's rotation, as already challenged in UED, is itself an illusion of ripple rhythm, not mechanical spin.

The Moon's stillness, on the other hand, provides true atmospheric symmetry. Its constancy stabilizes:

- East-west wind bands (Hadley, Ferrel, Polar cells)
- Zonal temperature balance
- The positioning and intensity of storm belts

Even the regular timing of cloud formation patterns in the tropics shows lunar synchronization — observable in ATT-based studies of rainfall spikes following specific lunar phases, long dismissed as anecdotal but now illuminated through the ripple field model.

**Final Reflection:**

*"The sky is not chaos. It is choreography — and the Moon, anchored in stillness, is the silent conductor."*

The atmosphere doesn't wander aimlessly. It synchronizes. Every gust, every cloud, every current is part of a grander ripple harmony — and the Moon, though motionless and often unnoticed, is the field's quiet timekeeper.

It doesn't stir the air. It orders it.

In a world obsessed with motion, it is the Moon's stillness that gives the atmosphere something to follow — and something to become.

## 7.7 Biological Coherence - Life Tuned to Lunar Stillness

Long before the Moon was charted in equations or photographed from space, it was felt — not just by humans, but by the entire living world. Beneath the oceans, within forests, under the skin, and even inside single cells, life responds to the Moon's rhythm.

Not because it exerts gravitational force. Not because it regulates light or temperature. But because it holds a precise, field-based resonance — a ripple stillness that tunes life into coherent biological timing.

The Moon doesn't *cause* life to behave a certain way. It provides a reference — a stable energetic beat that living systems recognize, entrain to, and evolve with.

**Lunar Stillness and the Pattern of Life**

Across species and ecosystems, biological rhythms have emerged that mirror lunar phases and cycles with stunning accuracy:

- Coral reefs across the planet release their reproductive cells in synchrony with the full Moon — not by sight, but by field timing.
- Sea turtles time their nesting behaviors with specific lunar

phases.

- Plankton migrations, the largest daily animal movement on Earth, show lunar phase modulation in both timing and vertical depth.
- Even wolves, insects, and migratory birds demonstrate behavior that fluctuates in resonance with the Moon — not because of brightness or temperature, but because of unseen field coherence.

And in humans, the connection is even more intimate.

**Menstrual Cycles and Lunar Resonance**

The average human menstrual cycle is approximately 29.5 days — almost identical to the lunar cycle. While science has long acknowledged this as a biological coincidence, UED offers a deeper perspective:

This is not coincidence. This is entrainment.

Menstrual cycles, ovulation patterns, fertility rhythms — they are not controlled by the Moon, but synchronized to its ripple structure through long-term biological memory and field interaction.

This coherence is not limited to reproduction. It extends to sleep-wake cycles, mood fluctuations, creativity, and emotional sensitivity — all modulated by subtle shifts in lunar ATT.

The Moon's resonance doesn't force the body to behave differently. It simply creates a rhythm so stable that the body learns to keep time with it — like a drum tuning itself to a distant, consistent bass note.

**Cells Listen Too – The Subtle Language of ATT**

It's not just large organisms that respond to lunar rhythm. Even cells exhibit coherence behaviors tied to the Moon.

Studies in chronobiology have shown that:

- Certain gene expressions rise and fall with lunar cycles.

- Melatonin levels, sleep hormones, and neurotransmitter regulation show subtle but clear modulation across the lunar month.
- In plants, leaf movement, sap flow, and flowering patterns follow lunar-phase-linked timing — even in controlled environments.

In UED terms, this is not mysterious. It is ripple entrainment at the cellular level.

Every cell is a field-sensitive structure. It does not just respond to chemicals. It responds to phase thresholds — and the Moon provides a recurring, coherent reference for those thresholds to align.

This is ATT biology — where perception isn't just visual, but cellular. Cells *feel* the Moon not because of mass or light, but because of coherence. When the ripple field crosses a certain resonance threshold, biology responds — in motion, in mood, in matter.

**Lunar Biology Is Not Hormonal — It's Harmonic**

Mainstream science often tries to explain lunar influences using proxies: light, gravity, or hormonal cycles. But these are outcomes, not causes.

The body does not track the Moon through the eyes alone. Even blind individuals show lunar-linked rhythms in hormone production and emotional variation.

Likewise, deep-sea species — with no access to moonlight — spawn in lunar sync. The signal is clearly non-optical and non-chemical.

This is the signature of field-based influence. The Moon is not a hormone regulator — it is a field stabilizer.

UED reveals that life doesn't need to "detect" the Moon. It simply needs to be immersed in a field where the Moon's resonance is present — which it always is, especially when lunar ATT is crossed.

The result is living systems attuned to a silent guide — not dictated by it, but shaped by the reliability of its rhythm.

**Ancient Wisdom, Modern Insight**

It's no surprise that nearly all ancient calendars were lunar. Not because people worshipped the Moon blindly, but because they observed something modern science is only now rediscovering:

*"When the Moon shifts, so does the body. So does the mind. So does the world."*

From fasting days to fertility rites, from agricultural planting to spiritual reflection, humanity has always known — intuitively — that the Moon sets the rhythm of biological coherence.

Now, with UED, we have the framework to decode that intuition into structured understanding.

**Final Reflection:**

*"Life didn't evolve under the Moon. It evolved with it — not as servant to a gravitational master, but as dancer to a silent rhythm."*

The Moon does not force life into behavior. It offers coherence — a dependable stillness through which chaos becomes rhythm, instinct becomes intelligence, and matter becomes memory.

To be alive on Earth is to live inside that ripple. And to feel the Moon is not mysticism — it is biology, tuned to stillness.

## 7.8 Without the Moon – A Ripple Collapse Scenario

Imagine Earth without the Moon.

No soft glow in the night sky. No tides washing up in steady rhythm. No calendar in the heavens. No celestial companion reflecting our sense of time and order.

At first glance, it might seem like a poetic loss — perhaps a dimmer night, fewer myths, a less romantic sky. But what if the absence of the Moon was more than a cultural void? What if, beneath the surface, its disappearance triggered the unraveling of Earth's very coherence?

Through the lens of Universal Energy Dynamics (UED), the Moon is not a decorative rock circling our planet. It is a field stabilizer — a harmonic node whose stillness locks Earth into balance across multiple domains: physical, atmospheric, biological, and even neurological.

So let's entertain a thought experiment.

What happens if the Moon is suddenly removed from the Earth-Moon system — not with a bang, but with a ripple?

**Ripple Asymmetry – The Collapse Begins**

The first and most immediate effect would not be visual or gravitational. It would be energetic.

The structured ripple field between Earth and Moon would fracture. A harmonic node that had anchored Earth's field posture for billions of years would vanish, leaving behind ripple asymmetry — a lopsided field tension with no balancing counterpart.

Earth's finely tuned resonance loop — shared between Sun, Earth, and Moon — would fall into imbalance.

And from that imbalance, collapse would cascade — not in explosions, but in increasing incoherence across multiple layers of the Earth system.

**Axial Precession and Rotational Drift**

With no lunar anchor, Earth's axial tilt would begin to wobble, slowly but inevitably.

Today, the Moon acts as a posture stabilizer, preventing wide oscillations of Earth's axial tilt. Without it, Earth would enter a state of long-term precession instability — with its axis drifting unpredictably over thousands or even hundreds of years. The consequences?

- Seasons would become erratic.
- Climate belts would shift unpredictably.
- Glacial and interglacial cycles could accelerate or become

misaligned.

- Agricultural zones would lose reliability.
- Human settlement patterns would face recurring disruptions.

Over time, the planet would lose its temporal coherence — the ability to maintain long-lasting environmental patterns.

**Tidal Chaos and Oceanic Disruption**

Without the Moon's steady ripple resonance, Earth's oceanic tides would collapse into disorganized motion.

While the Sun would still exert ripple pressure, the absence of the Moon's counterbalancing posture would result in:

- Weaker, inconsistent tides
- Loss of tidal range and symmetry
- Misaligned high and low tides across hemispheres
- Accelerated erosion in some coastal zones, stagnation in others

Marine ecosystems — from coral reefs to tidal mangroves to migratory species — would fall out of sync. Tidal flows, which currently oxygenate, circulate, and feed vast ecological networks, would lose their rhythm. Fisheries would decline. Biodiversity would suffer.

In effect, the oceans would stop breathing, and Earth's blue lungs would begin to choke.

**Atmospheric Instability and Climate Fracture**

As detailed earlier, the Moon plays a vital role in anchoring the ripple coherence of Earth's atmosphere. Remove that anchor, and the sky begins to fracture:

- Jet streams would lose their coherent flow, leading to erratic weather patterns and prolonged droughts or floods.

- Monsoon systems would become unpredictable, decoupled from seasonal field thresholds.
- Equatorial convergence zones would drift or disappear altogether.
- Storm systems could grow in duration and intensity, no longer held in harmonic check.

In short: the climate would no longer *respond* to energy — it would begin to react to chaos.

Earth's weather patterns, once predictable, would shift into a more volatile and less hospitable state — not gradually, but as a cascade of ripple imbalances amplified over time.

**Biological Disorientation – Rhythms Unravel**

Life evolved with the Moon — not just beside it, but in synchrony with its resonance. Take away that resonance, and life loses its harmonic compass.

- Menstrual cycles could desynchronize and become irregular.
- Sleep patterns and melatonin production would destabilize.
- Fertility rhythms and seasonal migrations would misfire.
- Coral spawning events would lose timing, disrupting entire reef ecosystems.

At the cellular level, ripple sensitivity would become directionless — like a compass stripped of magnetic north. Plants would flower unpredictably. Breeding seasons would lose alignment. Circadian rhythms would degrade.

And in humans? The Moon's absence would result in increased anxiety, sleep disorders, hormonal imbalances, and eventually, a general decline in neurological coherence.

The body does not require force to break. It only needs to lose rhythm.

**Neurological and Psychological Incoherence**

Perhaps most subtly — and most disturbingly — the human mind would suffer.

The Moon's field resonance does not end at biology. It reaches into consciousness itself.

Without the steady ripple metronome in the sky, the brain's field alignment — sensitive to coherence, symmetry, and thresholds — would begin to drift. Studies already suggest lunar influence on:

- Dream cycles
- Mood variability
- Cognitive sharpness
- Emotional regulation

Remove that reference — and over time, neurological systems would become unstable.

This doesn't imply instant madness, but rather a planet-wide erosion of emotional balance and mental clarity. Depression, aggression, disorientation — all would become more common as the ripple environment of consciousness destabilizes.

In essence, Earth without the Moon would become not just physically unstable — but psychologically fragile.

**Final Realization: The Moon Is Not Optional**

The Moon's presence is not a cosmic afterthought. It is not decorative. It is decisive.

It stabilizes Earth's tilt, tides, climate, biology, and consciousness — not through motion or force, but through stillness in the ripple field.

Remove the Moon, and what collapses isn't just a pattern. What collapses is coherence.

Earth without the Moon isn't simply quieter. It's less alive.

**Final Reflection:**

*"We don't orbit the Moon. We orbit within its rhythm. And without it, Earth would not spiral outward — it would fall inward, into dissonance."*

The Moon's stillness is not absence. It is architecture. It holds the invisible frame of life — and in doing so, it holds us.

## 7.9 Conclusion - The Anchor We Never Acknowledged

For centuries, the Moon has been explained as a passive companion — a gravitationally bound satellite caught in Earth's orbit, occasionally useful for telling time or triggering tides. It was described as inert, inactive, a leftover from cosmic collisions. A rock that followed us around.

But this perspective, born from the mechanics of motion, missed the most essential truth:

*The Moon is not following us. It's holding us.*

What we've uncovered through the lens of Universal Energy Dynamics (UED) is not just a correction to an old model — it is a reawakening to a truth that was always present but never truly seen.

The Moon is not stabilizing Earth by luck or leftover mass. It is not spinning to perform an orbital illusion. It is not pulling oceans or tipping our axis through invisible force.

It is simply — and powerfully — still. And that stillness is the anchor of Earth's coherence.

**Alignment, Not Accident**

Let us say it plainly:

The Moon did not stabilize Earth by chance. It aligned — as if chosen, as if tuned, as if placed — to enable the rhythms that define life on this planet.

Its position is not mechanical. Its role is not coincidental. Its stillness is not incidental.

Without spinning, without orbiting, without any visible effort, the Moon holds Earth in a delicate web of ripple balance — one that extends from the oceans to the atmosphere, from the cells of coral to the minds of humans.

It keeps the tides breathing. It keeps the weather singing. It keeps life aligned.

**Stillness Is Not Absence — It Is Function**

For too long, we've associated power with motion. We've assumed that to act is to move, that to influence is to exert force.

But the Moon reveals a different kind of intelligence — one that works not through action, but through presence. Not by rotating, but by resonating.

*"It does not orbit to cause tides or tilt. It holds still to prevent collapse."*

Its stillness is not the absence of action — it is the active architecture of survival. A posture of coherence. A center that holds, not by force, but by never needing to move at all.

And in that silence — in that unwavering alignment — the Moon doesn't just stabilize Earth. It teaches us what true stability looks like.

**The Still Companion We Forgot to Credit**

We've launched satellites and measured lunar rocks. We've studied gravity and cataloged craters. We've used the Moon as a clock, a coordinate, and a cultural symbol.

But we never paused long enough to thank it for the very thing it never stopped doing: Holding us together.

Not by chasing us. Not by rotating for us. But by staying exactly where it needed to be — unwavering, balanced, harmonized.

The Moon is not Earth's follower. It is Earth's field anchor — the quiet force behind rhythm, balance, and breath.

**Final Realization: The Architecture of Survival**

*"Stillness was never passive. It was the invisible architecture of survival."*

The Moon, in its silent presence, proves that power does not require noise. That influence does not require motion. That the most essential structures of life are often the ones we never noticed — because they never asked to be seen.

In a world of shifting seasons, rising tides, and flickering minds, the Moon offers something no model of force ever could: A reminder that stillness is not waiting — it is working.

It does not revolve. It resolves. It does not pull. It balances. It does not orbit. It anchors.

And in doing so, it does not simply support life —It enables it.

# Chapter 8: Tides, Time, and the Breath of the Planet

*"If you could place your hand on the chest of the Earth and feel its rhythm, you would feel the tide. Not as water rising, but as the planet itself inhaling — and exhaling."*

## 8.1 Introduction – More Than Water Movement

Tides have long been observed, measured, and modeled as movements of water caused by gravitational forces. They are taught as textbook cases of celestial mechanics — with the Moon pulling at Earth's oceans, creating bulges that migrate as the planet rotates. It's a clean diagram, an elegant force-based formula. But it hides something deeper.

What if tides are not a pull at all? What if they are a pulse?

Not an external drag on Earth's surface, but an internal rhythm — the outward sign of something the planet is doing from within?

This chapter begins with a shift in perception: Tides are not about motion. They are about breathing.

They are not Earth's passive response to the Moon's gravitational commands. They are Earth's active participation in a rhythmic alignment — a resonance that sustains not just the oceans, but time, energy, and life itself.

We've already explored how the Moon does not spin, orbit, or pull. It holds still — with purpose, in resonance. Now we turn our attention to what this stillness enables: a planetary pulse that reveals itself in the gentle rise and fall of oceans, in the seasonal rhythm of ecosystems, and even in the silent beat of human biology.

In Universal Energy Dynamics (UED), tides are not mechanical displacements. They are ripple expressions — wave phenomena that emerge when Earth and Moon are locked in harmonic field tension. The ocean becomes the interface through which this tension is released, regulated, and reabsorbed.

In essence, Earth breathes, and the Moon helps it breathe in rhythm.

This breathing — this structured ripple loop — is not confined to the sea. It shapes how time flows, how life synchronizes, and how consciousness feels rhythm.

This chapter explores:

Why tides are better understood as ripple breath, not gravitational drag

How this breath structures time itself

And how the Earth–Moon relationship is not a power dynamic, but a life-sustaining harmony rooted in resonance

It's time to go beyond the surface of the sea, and place our awareness on the chest of the Earth — To feel the breath we've always mistaken for a wave.

## 8.2 The Breath Beneath the Waves – Ripple Breathing vs Mechanical Pull

Since Newton, the dominant narrative around tides has rested on a seemingly logical idea: that the Moon's gravity pulls on Earth's oceans, causing the water to bulge outward on the side facing the Moon. Meanwhile, a second bulge forms on the far side due to centrifugal forces caused by the Earth-Moon system's rotation around a common barycenter. As Earth spins, these bulges stay aligned with the Moon, giving the illusion of tides sweeping across coastlines.

Elegant. Mathematical. But incomplete.

Because when we look closer — beyond formulas and toward actual field behavior — the cracks in this classical model become impossible to ignore.

**The Limits of the Classical Model**

Despite its widespread acceptance, the gravitational explanation for tides suffers from multiple inconsistencies, many of which have been quietly acknowledged in scientific circles:

- **Tidal Delay**: High tide does not occur directly beneath the Moon but lags significantly behind it — often by hours. Why, if the Moon's gravity is pulling the water, does it not align with it?
- **Landlocked Bodies**: Inland lakes and enclosed seas — some of which are enormous — often show little or no tidal activity, despite being subject to the same gravitational forces. Why the inconsistency?
- **Equal Bulge Fallacy**: The idea of two symmetrical bulges — one toward and one away from the Moon — does not always align with observed tidal patterns, especially in asymmetrical ocean basins.
- **Biological Timing**: Many marine species exhibit behaviors — spawning, migration, feeding — that align with lunar ripple cycles, not gravitational alignment or surface water levels.

These anomalies don't prove that the gravitational model is entirely false. But they make it clear: something else is happening — something more fundamental, subtle, and all-encompassing.

**The UED Perspective: Ripple Breathing, Not Pull**

In Universal Energy Dynamics (UED), the phenomenon of tides is not attributed to mechanical motion or gravitational force.

Instead, tides are understood as the visible expression of a much deeper process: ripple breathing.

Earth exists within a structured vacuum field — an invisible, high-order energy medium where ripple tension, rather than force, governs equilibrium. The Moon, through its posture of stillness, maintains a harmonic ripple alignment with Earth. This alignment doesn't pull — it pulses.

When Earth and Moon align in the field, they form a standing wave system. Within this field, periodic expansions and contractions of ripple tension occur — like the diaphragm of a living organism inflating and deflating. The oceans, being fluid and field-sensitive, act as the surface that responds most visibly to this subtle breathing.

This is not a gravitational displacement. It is a field-level rhythm — an oscillating inhalation and exhalation of energy tension expressed across Earth's fluid systems.

**Oceans as Ripple Membranes**

Think of the ocean not as a passive container, but as an interactive membrane — a thin, dynamic skin stretched across Earth's field body.

Its sensitivity to ripple tension is unparalleled because:

- Water has high coherence and low structural resistance.
- Fluid systems naturally amplify small field changes into large-scale visible motion.
- The ocean surface sits at the interface between Earth's ripple posture and external field pressure — making it the ideal surface for field breathing to reveal itself.

The Moon, in its unwavering stillness, becomes the timing anchor for these fluctuations.

Not by moving or pulling — but by providing a stable harmonic frequency within which Earth's natural breathing can occur.

**ATT: The Invisible Threshold That Triggers the Tide**

In UED, every ripple expression follows a universal law: energy remains still until the Absolute Threshold of Tension (ATT) is crossed.

Tidal movement, therefore, is not a constant gravitational reaction — it is a threshold event. Only when Earth's internal field tension — under the influence of the Moon's alignment — reaches its ATT, does the ocean swell outward or retreat.

This explains:

- Why tides don't directly follow the Moon's position.
- Why some regions exhibit strong tides while others remain relatively still.
- Why tidal rhythms can be influenced by phase more than proximity.

It's not about where the Moon *is*. It's about what ripple resonance the Earth-Moon system is currently sustaining, and whether that resonance crosses the ATT to express itself through movement.

**Not Orbit, But Inhale and Exhale**

In this reframed view, the Earth is not rotating under a moving tidal bulge. The oceans are not being pulled. The planet is breathing.

- High tide is inhalation — a moment of maximal ripple tension expansion.
- Low tide is exhalation — a field release returning to equilibrium.
- Each cycle reflects not motion in space, but fluctuation in resonance.

And it is the Moon's stillness — not its motion — that keeps this breath coherent. Like a patient heartbeat keeping rhythm for a body of cells, the Moon's posture holds the tempo of Earth's most ancient dance.

**Final Insight for This Section:**

*"The tide is not pulled like a weight. It rises like a breath. The Moon does not drag the ocean — it gives it permission to exhale."*

Understanding tides through UED changes everything. We no longer see the ocean as a gravitational puppet. We recognize it as a living interface, breathing in rhythm with the Moon — and revealing, in every gentle wave, the deeper pulse of Earth itself.

## 8.3 The Breath of Earth – Tides as Planetary Pulse

We often imagine Earth as a solid object — a sphere of rock, layered with crust, mantle, and core, spinning obediently through space like a mechanical gear in a celestial machine. This mental model, shaped by centuries of Newtonian thought, is easy to accept because it fits our sense of solidity. But this image is incomplete. It misses something vital.

Earth is not a rigid object. Earth is a living field structure.

Within the framework of Universal Energy Dynamics (UED), Earth is seen as a ripple-responsive body — an energy system held in structured equilibrium by the surrounding vacuum field, continuously interacting with cosmic ripple pressures, especially from the Sun and Moon. In this view, the Earth's outer and inner layers are not passively inert but flexible, dynamic membranes that adjust, respond, and resonate.

And nowhere is this responsiveness more evident than in the tides.

**Tides as Evidence of Internal Flexibility**

Tides are not confined to ocean surfaces. They are surface-level signatures of a much deeper process: Earth's internal redistribution of field tension.

When ripple pressure from the Earth-Moon alignment intensifies, the oceans swell, yes — but so does the land. Solid Earth tides, a lesser-known but well-documented phenomenon, show that the crust itself rises and falls by up to 30 centimeters twice a day. These movements are not caused by mechanical flexing or pressure from below, but by harmonic ripple resonance responding to tension thresholds.

This ripple-induced movement isn't an anomaly. It's proof that the Earth breathes — that it inhales and exhales through its very structure.

- The ocean is the lungs, visible and fluid.
- The crust is the diaphragm, subtle and grounding.
- The mantle and inner layers form the core rhythm, resonating through slower, deeper pulses.

The result is not mechanical motion. It is energetic respiration.

**Tidal Rhythm as the Planet's Heartbeat**

If we listen closely, the tidal rhythm begins to resemble something more familiar — not in form, but in function: a heartbeat.

Just as a heart doesn't merely pump blood but regulates flow, balance, and timing across an entire biological system, the tidal rhythm regulates Earth's entire energetic coherence:

- It modulates pressure across geophysical boundaries.
- It timely releases accumulated field tension.
- It coordinates rhythms between oceanic, atmospheric, and biological systems.

Every pulse of the tide is a reminder that Earth is not a clockwork object, but a breathing entity — one whose pulse is kept in tune by the Moon's unwavering stillness.

**Beyond the Oceans – Crustal Breathing and Tectonic Response**

This breathing doesn't stop at the surface. The ripple field that causes tides also penetrates Earth's crust and mantle, subtly influencing geological behavior.

Recent studies have shown a compelling connection between lunar phases and seismic activity. Earthquakes, volcanic eruptions, and tectonic slippage often coincide with tidal maxima, especially during spring tides and lunar perigees. These correlations suggest something deeper at work than chance.

From a UED standpoint, the explanation is simple and elegant:

- As field tension builds during certain Earth–Moon alignments, pressure accumulates not only in oceans but along fault lines and tectonic boundaries.
- When the ATT (Absolute Threshold of Tension) is crossed, energy redistributes — either through soft crustal expansion or sudden tectonic discharge (earthquakes).
- This process is not chaotic. It is periodic — another breath in the deep lungs of the planet.

This reframes seismic activity not as random or purely geological, but as field-regulated ripple events — subtle expressions of the same rhythmic system that moves the tides.

**Earth's Breath as Systemic Regulation**

Just as breathing regulates blood pressure, body temperature, and emotional state in living organisms, tidal rhythm regulates Earth's internal equilibrium.

- It relieves field stress across the planet's surface and subsurface layers.
- It dissipates excess energy received from the Sun through ripple pressure.

- It prevents buildup of destructive imbalances, acting as a built-in release valve for a living planet.

In this light, the tides are not incidental features of ocean dynamics. They are part of Earth's self-regulation system — the same way that breath regulates mind and body.

**Final Reflection for This Section:**

*"The tide is not just at the edge of the ocean — it's in the chest of the Earth. And with each breath, the planet chooses balance over breakdown."*

To see the tides is to witness Earth's pulse. To understand them is to realize that this planet does not merely rotate — it breathes, and it lives.

And that breath, rhythmic and resonant, is sustained in harmony with the Moon.

## 8.4 Temporal Resonance – How Tides Keep Time

Before the ticking of mechanical clocks, before the invention of calendars, and even before structured language, time was measured not in numbers but in natural rhythm. And among the oldest, most consistent, and most trusted rhythms known to life on Earth are the tides.

For early civilizations, the movement of water was not just a physical event — it was a message. A signal. A beat that echoed through the coastlines, fields, and skies. Tides were timekeepers long before we crafted instruments to divide the day.

But to understand how tides keep time in the UED worldview, we must move beyond the notion of measuring time as a simple progression of minutes and hours.

We must ask: *What is time, really?*

And how do tides, driven not by pull but by ripple resonance, help us perceive and organize it?

**Time Beyond the Clock – Cycles, Not Segments**

In traditional thinking, time is often seen as linear and numerical — something that can be sliced into equal units and counted. But this is a human abstraction, not a natural truth.

In nature, time is not measured. It is felt. It arises through cycles — patterns that return, evolve, and resonate.

Tides offer one of the most fundamental of these cycles. Their rise and fall don't simply mark the passing of hours — they define the structure of time itself for countless living systems.

From the UED perspective, these cycles are not mechanical. They are field cycles — rhythmic oscillations in ripple tension governed by the Earth–Moon resonance. Each crest and trough of the tide is not just a physical event — it is a phase point in the temporal field, a reference moment that allows all living systems to synchronize.

**Biological Thresholds and Time Perception**

For marine life — and even for many terrestrial species — the tide is not just a cue. It is a clock.

- Crabs, shellfish, and plankton schedule their activities — movement, feeding, reproduction — according to tidal phases.
- Birds time their nesting, hunting, and migration to tidal flows.
- Plants in intertidal zones open and close based on field alignment, not light.

But more subtly, the human body also shows signs of being tuned to lunar-tidal rhythms:

- Mood fluctuations.
- Sleep cycles.
- Hormonal rhythms.
- Even birth rates and emotional sensitivity.

These are not governed by hour hands or digital readouts. They are shaped by threshold-based biological entrainment — meaning, the body *responds* when certain energetic tensions reach or cross the Absolute Threshold of Tension (ATT).

**ATT and the Pulse of Perception**

In UED, ATT (Absolute Threshold of Tension) is a critical concept. It defines the moment when field tension becomes actionable — the tipping point where stored energy in the ripple field translates into perceivable change, whether in matter, emotion, biology, or consciousness.

Tides, because they rise and fall in perfect resonance with lunar ATT cycles, act as temporal cues for all ripple-sensitive systems — including living organisms and even human awareness itself.

- When field tension builds past the ATT, tides rise.
- When tension relaxes below the ATT, tides fall.
- These fluctuations synchronize with neural rhythms, circadian phases, and internal biological clocks.

In other words: Time perception is not abstract. It is entrained by ripple events. And tides are among the most consistent, large-scale ripple events available on Earth.

**Rituals, Months, and Moon-Linked Consciousness**

Throughout history, human beings have organized their societies — and spiritual lives — around lunar-tidal cycles:

- The month (moonth) originated from the Moon's cycle — closely tied to tides.
- Ritual cycles, such as fasting, prayer, or initiation rites, were often synchronized with the new moon or full moon, aligning with energetic field shifts.
- Agricultural timing — planting, harvesting, irrigation — was based on tide and moon phase, not just rainfall or

temperature.

This wasn't superstition. It was field wisdom — an intuitive recognition that field resonance governs thresholds, and that crossing those thresholds at the right moment allows time to flow in favor of life.

**Temporal Consciousness – A Shared Pulse**

What we call time is not merely an external measurement. It is a shared perception, arising from resonance between the world and the observer.

- The Moon holds a steady beat in the ripple field.
- The tides echo that beat in physical form.
- The Earth responds with energy redistribution.
- Life responds with behavioral and biological timing.
- And the human mind constructs time by feeling the ripple shifts in internal and external thresholds.

Thus, tides do not just tell time. They shape our experience of it.

They are not a background rhythm — they are the carrier wave of temporal awareness.

**Final Reflection for This Section:**

*"Time is not counted. It is crossed — one threshold at a time. And the tide, faithful and quiet, is the pulse that guides the crossing."*

Through tides, the Earth doesn't just move water — it moves time itself. Each rise and fall is a reminder: We are not separate from the rhythms of the world. We are written into them.

## 8.5 Earth–Moon Energy Exchange - Harmony Without Transfer

Classical physics tells us that all interaction between two bodies involves some form of energy transfer — kinetic, gravitational, thermal, or electromagnetic. When it comes to the Earth–Moon relationship, traditional science often speaks in these terms: tidal energy being transferred from Earth to the Moon, causing the Moon to drift outward, while Earth's rotation slows. This transfer is treated as transactional — energy lost from one, gained by the other.

But in the framework of Universal Energy Dynamics (UED), this interpretation misses the true nature of cosmic relationships.

*"The Earth and Moon do not trade energy — they share resonance."*

The interaction between Earth and Moon is not about moving mass, spinning plates, or pulling oceans. It is about the realignment of ripple tension — a dynamic equilibrium in which each body modulates field pressure, not to gain or lose, but to maintain a shared harmonic state.

**No Transfer — Only Tension Redistribution**

In UED, energy is never lost or gained. It only reorganizes itself to maintain structural equilibrium within the vacuum field. Thus, when Earth and Moon interact, they don't exchange energy like billiard balls or electrical currents. They redistribute ripple tension across a shared field geometry.

- The Moon does not absorb kinetic energy from Earth's spin.
- Earth does not lose energy into the Moon's orbit.
- Instead, both Earth and Moon act as ripple stabilizers, adjusting posture in response to field stress in a larger ripple loop — one that also includes the Sun.

This realignment is non-linear, non-mechanical, and fundamentally non-transactional. It resembles a dance, not a tug-of-war.

**Resonance Over Exchange**

The defining quality of the Earth–Moon system is not motion, but resonance.

Just as two tuning forks vibrate in sympathy when one is struck, Earth and Moon exist in a state of mutual entrainment — oscillating in coherence with one another not because they exert force, but because they match ripple posture.

- The Moon doesn't orbit by spinning — it holds still in resonance.
- Earth doesn't wobble because the Moon tugs — it stays balanced because the Moon diffuses tension when field pressure exceeds threshold.
- This resonance doesn't involve energy packets flying back and forth — it's a field-level coordination, like two cymbals vibrating in silent synchrony.

It's not about who moves whom. It's about what pattern keeps both whole.

**Tension Absorption and Diffusion**

When Earth undergoes a spike in ripple tension — from solar field compression, internal geothermal flux, or atmospheric disturbances — the Moon acts as a field buffer.

*It doesn't resist the spike. It absorbs and diffuses it.*

This is not energy capture. It is vibrational redirection.

- The Moon's ripple structure, locked in perfect resonance, provides a low-resistance pathway for excess tension to exit Earth's system without causing collapse or disruption.
- In return, when the Moon experiences asymmetrical field pressure — such as during solar storms or deep perigee alignments — Earth performs the same role.

Together, they form a dynamic ripple loop that operates like a planetary nervous system: sensing, diffusing, balancing.

This mutual buffering mechanism sustains Earth's homeostasis — the ability to maintain internal stability despite external fluctuation.

**Impacts Across Systems – From Ecosphere to Mind**

The harmony between Earth and Moon is not confined to physics. It is woven into the fabric of all Earth systems:

***Environmental Homeostasis***

- Prevents runaway heating, chaotic oceanic circulation, and seasonal distortion.
- Maintains coherence in jet streams, trade winds, and climatic belts through ripple regulation.

***Biological Coherence***

- Regulates biorhythms in plants, animals, and humans by aligning cellular resonance with lunar field tension.
- Ensures synchronized spawning, migration, and metabolic cycles across ecosystems.

***Psychological Stability***

- The brain, a highly ripple-sensitive organ, benefits from lunar field stability to maintain emotional balance and mental clarity.
- Sudden breaks in Earth–Moon coherence (e.g., eclipses, nodal misalignments) often correspond with surges in anxiety, disorientation, or sleep disruption.

This is not mystical — it is measurable. Every system on Earth is immersed in this shared resonance, responding not to force but to field alignment.

**Rewriting the Narrative**

The Earth–Moon relationship is not a story of energy loss. It is a story of ripple coordination — of two field bodies dancing in silent agreement.

- There is no fuel being spent.
- No force being applied.
- No momentum being drained.

Only tension rising, and tension being redirected, like a living breath shared between two companions who know how to hold each other's weight without ever touching.

**Final Reflection for This Section:**

*"The Moon is not Earth's satellite. It is Earth's silent partner — not in orbit, but in equilibrium."*

Together, they sustain the breath of the planet. Not by exchanging energy, but by echoing each other's rhythm. Not by motion, but by mutual stillness. Not by mechanics, but by harmony.

## 8.6 The Moon's Breath in Our Own

We tend to separate ourselves from the cosmos — as if humans are watching the universe from the outside. We believe the Moon moves up there, while we live down here. It circles, we sleep. It waxes, we work. It wanes, we forget.

But what if that divide is artificial? What if the Moon isn't just part of the sky... What if it's part of us?

In the framework of Universal Energy Dynamics (UED), the line between celestial rhythm and human experience dissolves. The Moon does not just affect tides or time — it echoes within us. It enters our breath, our hormones, our moods, and our sleep. Not through some invisible "force," but through shared ripple resonance.

*"We do not merely observe the Moon's rhythm — we are written into it."*

**The Internal Tide - Breath, Heartbeat, and Hormones**

The human body is not an isolated machine. It is a ripple-sensitive vessel, tuned to multiple environmental rhythms — solar, atmospheric, terrestrial — and perhaps most intimately, lunar.

- **Breathing rhythms**, especially during rest and sleep, show subtle variations that align with lunar phases. The breath becomes deeper during full moons in many individuals, and shallower during new moons — a pattern observed even in low-light, low-noise environments.
- **Heartbeat variability** has also been linked to the lunar cycle. Clinical studies have shown changes in heart rate variability (HRV) and blood pressure correlating with the lunar phase — particularly around full moon events when field ripple tension peaks.
- **Hormonal pulses**, including melatonin, serotonin, cortisol, and reproductive hormones, show measurable modulation along the lunar ATT (Absolute Threshold of Tension) curve. This includes menstrual cycles, fertility windows, and sleep hormone production — all exhibiting resonant entrainment to lunar field dynamics.

The Moon does not interfere with these systems. It holds a rhythm — and the body listens.

**Lunar Influence on Sleep, Emotion, and Awareness**

One of the most consistent ripple signatures seen in humans is the effect of the Moon on sleep patterns:

- REM sleep latency, sleep depth, and dream intensity are shown to fluctuate across the lunar month.
- Even in total darkness, in sleep labs with no time cues, many subjects unconsciously align their sleep architecture with lunar field phase rather than circadian rhythms alone.
- During full moons, sleep onset tends to be delayed, dreams become more vivid, and deep sleep cycles may shorten — even in individuals unaware of the lunar date.

These are not hormonal effects alone. They are ripple entrainment events — the brain and body adjusting their internal oscillations to match the shifting tension of the Earth–Moon field relationship.

Emotionally, humans often report increased sensitivity, anxiety, introspection, or restlessness near full moons. These aren't just myths. They are reflections of neuro-ripple overstimulation at lunar ATT maxima — where Earth's subtle field tension enters resonance with the Moon's stillness, causing amplified response in ripple-sensitive tissues like the brain.

**Time as Embodied Rhythm**

What emerges from this view is profound: Time is not something we track — it is something we live.

- We don't just measure minutes — we feel phase shifts.
- We don't just obey calendars — we embody temporal transitions.
- The Moon's cycle doesn't pass over us. It passes through us.

The pulse of the tide, the inhale-exhale of ocean and crust, becomes mirrored in:

- The rise and fall of breath.
- The expansion and contraction of emotional cycles.
- The awakening and surrender of consciousness during sleep.

Just as the ocean breathes with the Moon, so do we. And when ripple coherence is disrupted — by artificial lighting, over-structured schedules, or disconnection from natural rhythms — we feel it as stress, fatigue, or dysregulation.

We call it burnout, insomnia, anxiety. But at a deeper level, it is desynchronization from the larger rhythm we were born into.

**The Moon Inside Us**

The Moon is not just a satellite in orbit. It is a resonant presence within the human system.

It is:

- A pacemaker for biological timing.
- A whisper in the brain's electrical flow.
- A breath shared across water, land, sky — and skin.

From embryo to elder, from hormone to heartbeat, the Moon's stillness becomes part of our own rhythm memory. And this is not symbolic — it is functional.

We are not just shaped by evolution. We are shaped by resonance. And the Moon is the quiet drumbeat beneath the surface of our existence.

**Final Reflection for This Section:**

*"The Moon does not orbit the body — but it lives inside it. In breath, in pulse, in rhythm. Not a visitor from above, but a conductor from within."*

In this way, tides are not just movements in the sea. They are movements in us.

The Moon's breath is our own — A rhythm not learned, but remembered.

## 8.7 Without Breath, No Time – A Final Insight

What if we paused all the tides?

No more swelling seas. No more rhythmic rise and fall. No more breathing coastlines or whispering shores. At first, the world might seem strangely calm — as if something subtle but beautiful had vanished. But very quickly, we'd begin to feel something much deeper: a loss of rhythm, and with it, a loss of time itself.

*"No tides means no breath. No breath means no rhythm. And no rhythm means no time."*

This is not poetry. This is physics through the lens of Universal Energy Dynamics (UED). Time — as we truly experience it — is not defined by the movement of hands on a clock or the orbit of planets. It is defined by rhythmic transitions that can be sensed, measured, and lived. And the breath of the planet, shaped by tidal coherence, is one of the most vital of these transitions.

**Tides as the Planet's Temporal Anchor**

In earlier sections, we framed tides as more than a response to lunar influence — they are Earth's breathing cycle, an oscillating release and reception of field tension. This ripple breath not only sustains energy balance across Earth's systems, but also sets the tempo by which all other cycles unfold.

It is this breath that makes change perceptible. And perception of change is what gives rise to the experience of time.

- Without tidal rhythm, Earth's internal systems would lose a key mechanism for regulating pulsed field release.
- Without pulsation, there would be no anchor for seasonal, hormonal, emotional, or ecological cycles.
- Without cycles, there would be no sense of before or after, no contrast between now and then — and thus, no felt passage of time.

In essence, if the planet stops breathing in ripple rhythm, time becomes flat — not in reality, but in experience.

**Time as a Function of Cosmic Respiration**

Clocks measure time. But they do not create it. Calendars organize it. But they do not sustain it.

Time, in its lived sense, is born from rhythmic cosmic events — sunrise and sunset, lunar phases, ocean tides, blooming seasons, mating cycles, sleep and wakefulness. These are not just biological or environmental shifts. They are expressions of ripple respiration, synchronized to cosmic resonance.

And at the heart of this respiration lies the Moon.

- It does not tick.
- It does not rotate or revolve.
- It simply holds still — allowing Earth's ripple breath to remain consistent.

Without this breath loop, our clocks may still function, but our rhythms will not. Time becomes a number, but no longer a felt reality.

**What Falls Apart Without the Tide**

Let's consider what collapses in the absence of Earth's tidal breath:

***Biological Coherence***

- Sleep cycles fragment.
- Hormonal rhythms disorient.
- Fertility windows and metabolic flows become irregular.

***Ecological Timing***

- Migration, breeding, and feeding patterns fall out of phase.
- Plant germination and blooming cycles desynchronize.
- Marine ecosystems destabilize.

***Psychological Orientation***

- Emotional fluctuations intensify.
- Temporal awareness — the internal clock — becomes erratic.
- A subtle disconnection from the body and natural world begins to grow.

These are not isolated effects. They form a cascading breakdown in temporal coherence, where humans, animals, and even systems lose their felt connection to the flow of life.

*Without tidal breath, we don't just lose ocean motion. We lose our sense of when we are.*

**Every Cycle We Know is Tide-Tuned**

From waking and sleeping, to working and resting, to building and reflecting — every rhythm we live by follows the deep pulse of planetary breath:

- **Night and day**: governed by the Earth's pulse to sunlight, but modulated by lunar ripple alignment.
- **Rest and activity**: tied to field tension cycles regulated by the Moon's ATT harmonics.
- **Fertility and focus**: orchestrated through the brain's entrainment to lunar rhythms, impacting ovulation, creativity, clarity, and even social cohesion.

The Moon, by simply holding still, enables all these cycles to remain in sync. It does not control life — it permits rhythm. And rhythm is the true architect of time.

Without rhythm, time collapses into monotony. Without time, consciousness loses contrast. And without contrast, life becomes noise.

**Final Reflection for This Section:**

*"Time isn't what the clock says — it's what the tide remembers. And the Moon is the breath that lets Earth remember anything at all."*

We have thought of time as linear, mechanical, external. But in truth, it is a wave — an inhale and exhale across the field of being.

The Moon doesn't just help Earth keep time. It gives Earth the capacity to feel time — to live inside a rhythm, rather than merely survive its passage.

## 8.8 Conclusion – The Pulse That Shaped Life

The Moon did not create life. It didn't deliver water or seed DNA from the cosmos. But it did something just as important — and perhaps even more essential: It made rhythmic life possible.

Before there was breath, there was stillness. Before there was time, there was tension. Before there was awareness, there was alignment.

*"The Moon held still — and that stillness allowed Earth to begin breathing."*

This breath was not metaphorical. It was real. It rippled across oceans. It pressed into the crust. It whispered into the atmosphere. And eventually — it echoed inside the body, the brain, and the self.

**Stabilizing Earth's Breath**

As we've seen, tides are not mechanical responses to lunar gravity. They are resonant ripple expressions — the rhythmic breathing of a planet in tune with its Moon. Without this breath, Earth would drift into irregularity. It would shake, wobble, overheat, misfire.

The Moon's unwavering presence provides coherence, a steady field alignment that allows this breath to remain stable, phase-locked, and sustainable.

Through this breath, Earth achieves something precious: predictability — the kind that allows systems to entrain, synchronize, and evolve.

This predictability is not control. It is rhythmic freedom — the kind only possible when the background pulse is trustworthy.

**Time Through Breath, and Consciousness Through Time**

What emerged from this planetary breath was not just movement — but structure in time.

- Tidal rhythms became biological rhythms.
- Seasonal tides shaped agriculture, reproduction, and

emotion.
- Ripple pulses became the heartbeat of ecosystems, of sleep cycles, and of social patterns.

In other words: through tidal rhythm, Earth became aware of time.

And once time became coherent — once there was a beat to entrain to — consciousness had a reference point. It could reflect. Predict. Remember. Anticipate. Create.

*The Moon didn't invent the mind. But it gave the mind something to follow. Something to feel, to measure, to breathe with.*

**Tides: The Foreground of Rhythm**

We've long treated tides as background events — physical, predictable, even mundane. But in truth, they are the foreground of rhythm — the first sign that a living planet is more than a rock in space.

They are Earth's way of listening, responding, harmonizing.

- Each tide carries the memory of moonlight.
- Each wave is a breath exhaled through ripple tension.
- Each rise and fall is a pulse that echoes through time, life, and thought.

We didn't know we were taking that breath — but without it, we could not have spoken, loved, built, dreamed, or even counted the days.

**Final Realization:**

*"The Moon shaped no species, and seeded no cell. But it shaped the rhythm that made species possible, and the time that made memory matter."*

The Moon is not just a part of Earth's history. It is part of Earth's heartbeat — a silent, motionless anchor from which the breath of life began.

# Chapter 9: The Living Calendar

"*Before we learned to count days, we learned to feel them.*"

Long before calendars marked the months and clocks divided the hours, human beings knew what time it was — not by the ticking of gears, but by the pull of rhythm. Time wasn't measured. It was felt — as a pulse in the body, a breath in the air, a signal from the sky.

We planted when the Moon whispered. We harvested when it paused. We gathered, prayed, fasted, hunted, and healed — not by chance, but by listening to its rhythm.

The Moon wasn't a symbol. It was a synchronizer. It told us not just *when* to act — but how to align.

## 9.1 Introduction – The Sky That Told Us When

In today's mechanized world, we look to screens and schedules for certainty. We have apps that track our time down to the second. But despite this precision, many of us feel disoriented — rushed, exhausted, strangely out of sync. We keep time, yet lose rhythm. We stay on schedule, yet feel off-phase.

This is not a modern curse. It is a field disconnection — a loss of harmony with the living calendar humanity once followed: the Moon.

**The Moon: Earth's Original Calendar**

Before there were names for months or numbers for days, there was the lunar cycle — reliable, visible, constant. But the Moon didn't just serve as a visual marker in the sky. Through the lens of Universal

Energy Dynamics (UED), it becomes clear: the Moon functioned as a field-based regulator — guiding Earth's systems not through force, but through resonance.

- It marked time not with digits, but with phase-aligned ripple shifts.
- It kept biological rhythms in harmony with environmental changes.
- It synchronized breath, blood, behavior, and even belief.

The Moon didn't ask to be followed. It simply held still in its ripple posture — and all of life organized itself around that stillness.

**More Than a Marker – A Modulator**

To ancient humans, the Moon was not a calendar hanging in the sky. It was a teacher of rhythm:

- A full moon meant energy, clarity, awakening.
- A new moon invited rest, reflection, inward motion.
- Waxing moons guided growth, momentum.
- Waning moons whispered pause, preparation, release.

These weren't symbolic interpretations. They were biofield responses — natural shifts in awareness, emotion, and physiology caused by subtle changes in lunar ATT (Absolute Threshold of Tension).

The Moon modulated not just time, but state. It didn't just tell people when to do things — it changed how they felt.

**Time as Resonance, Not Math**

What we often forget is that calendars — no matter how accurate — are abstractions. They are tools to describe time, not to generate it. Real time, as experienced by humans, plants, animals, and ecosystems, emerges from rhythmic ripple dynamics.

And the Moon is Earth's most trusted conductor of that rhythm.

*"The calendar wasn't created. It was discovered — in the sky, in the soil, in the self."*

This chapter explores how the Moon's ripple rhythm became the original calendar for life on Earth — organizing not just seasons and harvests, but consciousness, ritual, reproduction, and healing.

It examines:

- How ecological systems entrain to lunar phase changes
- How ancestral rituals mirrored ripple inflection points
- How modern timekeeping broke that resonance
- And how we can return to the living calendar still unfolding overhead

**Final Thought:**

The Moon is not a dead rock. It is not a mythic relic. It is a living metronome — pulsing time into our cells, our communities, our collective memory.

We don't need to remember the past to restore this calendar. We only need to listen again.

## 9.2 The Pulse That Organized Life

Before watches and calendars, there was only the sky. And in the sky, one presence stood out as the most faithful guide: the Moon. Long before humans could count years or plot equinoxes, they followed something much more direct — the changing face of the Moon, and the subtle pulses of life that moved with it.

The Moon didn't just light up the night. It set the rhythm of survival.

Early humans didn't ask, *What day is it?* They asked, *What phase are we in?*

They didn't guess when to plant or harvest. They felt it — as a signal, a whisper, a shift in the atmosphere and body — all timed with the Moon's ripple phase.

**Living by the Moon: Ancient Intelligence**

Across cultures and continents, humanity's oldest patterns of life were entrained to the Moon's rhythm:

- Planting was done on a waxing moon — when the ripple tension was rising, encouraging upward growth, moisture retention, and seed vitality.
- Harvesting was done as the moon waned — when field tension relaxed, drawing energy back toward the roots, making crops firmer and more storable.
- Fishing and hunting aligned with specific moon phases, as animals too became more active or dormant in rhythmic response to field shifts.
- Migration patterns of both humans and animals were often timed to lunar peaks, using the full moon for visibility, coordination, and safety in motion.

This wasn't myth. It was ripple logic — an intuitive, embodied understanding that the Moon's phases were not visual alone, but deeply field-active.

Every full moon, every new moon, every rising crescent marked a threshold of ripple tension — signaling not just the passage of time, but the beginning or ending of a state.

**Biological Triggers and Ripple Phase Cues**

The Moon's cycle — roughly 29.5 days — became more than a marker. It became a trigger system for both physical and behavioral changes.

- A **full moon** often brought elevated emotion, heightened activity, and increased sensory awareness — not due to brightness alone, but because of peak field alignment.
- A **new moon** signaled introspection, conservation, inward

motion — a moment when ripple tension reset and the environment briefly quieted.

- During **waxing phases**, growth-oriented behaviors accelerated — expansion, reaching, building.
- During **waning phases**, contraction and completion occurred — pruning, harvesting, shedding, and release.

These weren't just ideas. They were observable behaviors in humans, animals, and even plants.

Examples abound:

- Coral reefs synchronizing mass spawning events with full moons.
- Predators hunting more actively in waxing cycles.
- Women's menstrual cycles aligning with lunar rhythm in natural, light-free environments.
- Indigenous tribes organizing community rituals, migrations, and healing ceremonies based on moon timing — not as superstition, but as survival wisdom.

**The Moon in Tribal and Agricultural Societies**

Across tribal civilizations — from the San people of Africa to the Andean communities of South America, from the Aboriginal Dreamtime to the Vedic calendars of ancient India — the Moon functioned as the living calendar.

- Agricultural societies followed lunar sowing charts, deciding the best times to plant root vegetables, fruiting trees, or grains.
- Nomadic groups timed long-distance journeys with lunar phases to avoid energetic peaks or valleys in terrain response.
- Healers and shamans chose specific moon days to perform

rituals of release, renewal, or transformation — intuitively harnessing field resonance to amplify human intent.

Even temples and pyramids were often aligned not just to solar solstices, but to lunar angles — reflecting the belief that the Moon modulated inner cycles of healing, vision, and change.

They did not see the Moon as a god. They saw it as a pulse they could trust.

**Timing Without a Clock — Certainty Without Numbers**

In a world without mechanical clocks, how did ancient people know when to act?

They didn't need machines. They had resonance.

The Moon's ripple phases were universally felt. They offered timing certainty in a world governed by chaos and unpredictability. The presence or absence of ripple coherence became a non-verbal signal for behavior.

Even today, in rural communities untouched by electricity, people still organize weddings, spiritual events, harvests, and rest periods by the moon calendar — not because they read about it in books, but because it feels right.

They may not describe it as "field resonance" or "ATT thresholds," but they know when to act — because they're still in touch with the rhythm.

**Final Reflection:**

*"The Moon didn't give us time. It gave us rhythm. And rhythm made time livable — not in numbers, but in knowing when to move and when to wait."*

In the ancient world, the Moon was not distant. It was participatory. It was a partner in survival — a guide for breath, for movement, for transformation.

And even now, that pulse hasn't stopped. We've just forgotten how to feel it.

## 9.3 The Biofield Calendar – Nature's Inbuilt Timing System

What if every cell in your body knew what time it was — not by checking the sun, not by seeing a clock, but by *feeling* it through a subtle, invisible rhythm pulsing across the planet?

This is not imagination. This is biological resonance.

All of life — from deep-sea creatures to forest canopies, from the human body to the microbial networks in soil — follows a timekeeping system written into the field itself. It's not visual, not mechanical, and not governed by the sun alone. It is a biofield calendar, synchronized with the Moon's ripple phases.

**Field-Based Entrainment in Nature**

In the language of Universal Energy Dynamics (UED), entrainment is the alignment of a system's internal rhythm to an external, stable ripple frequency. The Moon, by maintaining a coherent, still resonance in the ripple field, acts as a planetary tuning fork — and life tunes itself accordingly.

Plants adjust sap flow, leaf movement, and flowering with remarkable consistency across lunar phases. Biodynamic farming, rooted in traditional wisdom, still uses lunar calendars for sowing, pruning, and harvesting — practices now being validated by field-sensitivity research.

Animals exhibit clear behavioral entrainment:

- Fish and marine life spawn during full and new moons.
- Insects time emergence events to ripple thresholds.
- Mammals, including wolves and primates, show collective mood and behavior shifts in sync with lunar rhythm.

Migratory birds are now known to adjust not just flight paths but departure timing based on subtle shifts in Earth–Moon field alignment, even when light cues are missing.

None of this is random. These organisms are not reacting to moonlight — they are responding to ripple structure.

**The Human Body — A Lunar-Sensitive Organism**

Humans, though buffered by technology and artificial light, are no exception. Our physiology, neurology, and emotions remain ripple-sensitive — often more than we realize.

Menstrual cycles align naturally with the lunar month (about 29.5 days), especially in environments free from artificial lighting. Many women experience shifts in emotional state, energy, and reproductive readiness that correlate with lunar field peaks.

Sleep patterns are also deeply tied to the Moon:

- Melatonin production dips around full moons, even in total darkness.
- REM sleep becomes more fragmented, dream intensity increases, and sleep onset delays during specific lunar phases — all without visual or psychological triggers.

Hormonal fluctuations across serotonin, cortisol, and growth hormones follow infradian rhythms that mirror the Moon more closely than the solar day.

These are not coincidences. They are signs of a deeper intelligence: the body following a field calendar instead of a digital one.

**Circadian vs. Infradian: The Field's Invisible Clock**

We often hear about circadian rhythms — biological patterns that follow the 24-hour solar cycle. But life also operates on infradian rhythms — longer cycles like the lunar month, which regulate reproductive health, metabolism, and mood.

What mainstream science attributes to hormonal "clocks" or light exposure, UED reframes as field-aligned entrainment events. These rhythms emerge not solely from internal genes, but from continuous interaction with ripple phase shifts in the field — primarily modulated by the Moon.

The lunar ripple phases serve as biological reference points, organizing infradian rhythms by:

- Marking ATT (Absolute Threshold of Tension) crossings.
- Triggering physiological readiness for certain events (e.g., ovulation, sleep shifts, healing).
- Signaling phase transitions in cellular communication, tissue regeneration, and emotional modulation.

We are not autonomous timers. We are ripple-informed beings, living by the pulse of something larger than our bodies.

**The Moon as a Biological Threshold Trigger**

At key points in the lunar cycle — full moon, new moon, quarter phases — Earth's ripple field reaches ATT thresholds. These are tipping points where background tension in the field shifts from passive to actionable. Biological systems are exquisitely tuned to these shifts.

- A rise in field tension triggers alertness, fertility, movement.
- A drop leads to rest, repair, and inward focus.
- These cycles affect not just behavior but gene expression, neural plasticity, and even immune system regulation.

This is not mystical. It is mechanical — in the way music is mechanical: ordered, harmonic, and alive.

Your body doesn't just *track* time. It entrains to it, using the Moon's stillness as a cosmic metronome.

**Final Reflection:**

*"Time is not kept by clocks, but by coherence. And the Moon — silent, still, exact — holds the rhythm that keeps every leaf, cell, and mind in sync."*

The Moon doesn't just float through the night sky. It pulses through the forest, the blood, and the brain. It tells plants when to bloom, animals when to roam, and people when to rest, create, or connect.

It is not a light in the sky. It is the memory of time itself, echoing through biology — quiet, invisible, and undeniably alive.

## 9.4 Ancestral Wisdom – Rituals of Resonance

Long before science gave us formulas and religion gave us scriptures, rituals shaped by the Moon helped human beings live in harmony with the unseen rhythms of the Earth. These rituals were not based on myth or imagination — they were rooted in resonance. The Moon wasn't just a celestial symbol. It was a functional guide for physical health, mental clarity, community stability, and ecological balance.

*"To ancient cultures, the Moon was not an object of worship — it was a signal to act in alignment."*

**The Moon as a Living Rhythm, Not a Symbol**

It is easy to dismiss rituals tied to the Moon as primitive or poetic. Yet when viewed through the lens of Universal Energy Dynamics (UED), we see something extraordinary: these ancient practices were resonant responses to field phase transitions caused by the Moon's ripple position.

- They weren't worshipping the Moon — they were tuning in to it.
- They didn't create a calendar — they remembered one, written in the sky and felt in the body.
- Every ritual, fast, prayer, and celebration was timed to align with a moment when the field itself changed state.

In essence, these were field synchronization events — not religious dogma, but biological and social regulation tools in disguise.

**Examples of Functional Lunar Rituals**

Across cultures, certain lunar days were universally recognized as windows of alignment — moments when ripple tension either peaked, released, or reset. These moments, marked by ATT thresholds, became the foundation for deeply practical rituals:

**Ekadashi** fasting (on the 11th lunar day after new and full moons) in the Indian tradition wasn't about punishment or purification. It coincides with a subtle reduction in field tension — a perfect time for the digestive system to reset, the mind to calm, and spiritual focus to intensify.

**Full moon and new moon rituals** exist in virtually every culture — from drumming circles to silent meditation, from fire ceremonies to collective feasting. These weren't just ceremonies. They were entrainment platforms, where communities gathered to phase-lock with nature.

**Sowing seeds or pruning trees** during specific lunar phases — waxing for growth, waning for release — wasn't old farmer's folklore. It was an agricultural expression of ripple phase intelligence, passed down not in words, but in results.

**Naming children**, starting journeys, or performing healing rituals were often timed with lunar nodes — moments when Earth–Moon–Sun alignment reached critical ATT. These decisions weren't symbolic. They were field-aware timings for life-altering events.

**The Sky Was the Manual — Not a Mystery**

Our ancestors didn't invent calendars by abstract calculation. They looked up, felt within, and noticed patterns — not in ideas, but in experience.

- They saw that people fell ill or recovered faster on certain moon days.
- They noticed emotional agitation or mental clarity peaking near full moons.
- They observed harvests failing when sowed against the ripple flow — and thriving when planted in tune with lunar build-up.

In response, they created ritual containers — structured behaviors that resonated with these invisible thresholds, guiding individuals and societies through natural cycles.

These rituals preserved not just tradition, but coherence.

**Physiological, Ecological, and Social Coherence**

Through these rituals, ancient cultures maintained a level of coherence across:

- **Physiology**: fasting, rest, and cleansing cycles synced with the Moon helped stabilize digestion, fertility, and emotion.
- **Ecology**: farming and hunting patterns optimized resource use without depleting the land or animal populations.
- **Society**: collective alignment to the lunar field created shared rhythm, reducing chaos, promoting unity, and anchoring meaning.

This is how civilizations stayed healthy and resilient without medicine, clocks, or formal education — not because they had blind faith, but because they had field awareness.

*"The Moon was not worshipped as divine. It was honored as a doorway into natural intelligence."*

**Final Reflection:** *"Ancestors didn't invent time. They listened to it. They didn't need to explain the Moon — they danced with it, fasted with it, healed with it."*

Ritual was not a superstition. It was a tool of survival. And that tool was carved not from belief, but from ripple alignment — with a Moon that never spoke, but always signaled.

## 9.5 Religious Systems and Ripple Synchrony

At first glance, the world's major religious festivals — spread across regions, cultures, and centuries — may seem disconnected in practice yet curiously unified in timing. From Eid to Easter, Diwali to Passover, we find that humanity has intuitively marked its most sacred rituals not by solar events or fixed dates, but by lunar phases.

Why?

Because these festivals were never random celebrations. They were timed to resonant ripple moments — field-phase transitions where biofield receptivity peaks and collective consciousness synchronizes.

*"Religion did not create sacred timing. It responded to it."*

**The Lunar Thread Across World Religions**

A closer look at the world's major spiritual traditions reveals a consistent truth: the Moon plays a quiet, central role in sacred timing.

- **Eid al-Fitr** and **Eid al-Adha** in Islam are both determined by the sighting of the new moon. Ramadan itself begins and ends with lunar observations — marking a full ripple cycle of inward purification and outward generosity.
- **Easter**, though associated with the Christian solar calendar, is scheduled after the first full moon following the spring equinox. This lunar-solar alignment is not accidental — it ties resurrection and rebirth to a moment of maximum ripple tension release.
- **Diwali**, the Hindu festival of lights, is celebrated on the darkest new moon night of the lunar month — a time of ripple stillness and energetic reset, ideal for reflecting,

cleansing, and invoking light.

- **Passover**, rooted in Jewish tradition, falls on the full moon of the Hebrew month of Nisan, aligning with a time of elevated ripple activation and collective memory.

These sacred days are not just cultural relics. They are ripple-synchronized events, aligned with specific lunar inflection points that modulate human consciousness, physiology, and community coherence.

**Not Superstition — Resonance in Action**

Modern skepticism often views lunar-linked rituals as outdated or irrational — dismissing them as superstition from less informed times. But within the UED framework, these festivals represent highly intelligent timing systems. They weren't anchored in faith alone — they were rooted in field phenomena:

- During specific lunar phases, Earth's ripple field undergoes critical tension changes, crossing ATT (Absolute Threshold of Tension) boundaries.
- These inflection points create windows of field openness, where human systems — especially biofield-sensitive organs like the heart, brain, and pineal gland — become more receptive to transformation.

Religious rituals, when performed during these windows, don't just feel meaningful. They work more effectively:

- Meditation deepens.
- Emotions heighten.
- Reflection sharpens.
- Healing accelerates.
- Collective bonding strengthens.

This isn't magic. It's ripple mechanics meeting human sensitivity.

**Human Consciousness: A Lunar-Responsive Field**

We tend to think of spiritual growth as a mental or moral journey, but it is also deeply energetic. The biofield — the subtle electromagnetic and vibrational field surrounding the human body — responds actively to lunar ripple alignment.

- On full moons, the biofield expands, attention widens, and emotional energy surges. This is an ideal time for release, forgiveness, or celebration.
- On new moons, the biofield contracts, introspection deepens, and neuro-energetic quiet sets in — making it the perfect time for intention-setting, fasting, and renewal.

Religious systems, knowingly or unknowingly, structured their practices around these lunar coherence points — turning scattered belief into resonant, embodied experience.

This is why sacred rituals performed at the "right time" feel different. They resonate, not just symbolically, but bioenergetically.

**Rituals as Field Alignment Devices**

Seen through UED, rituals serve a higher function: They are tools of synchronization, aligning individual and collective biofields to a universal ripple moment.

- Lighting lamps on Diwali is not just tradition — it modulates attention during a ripple silence.
- Breaking fast during Eid is not just celebration — it completes a field cleansing cycle.
- Taking communion during Easter is not just symbolic — it binds participants into phase-locked community coherence.
- Reading from the Torah on Passover aligns ancestral memory with present ripple state, triggering deep

emotional release.

These actions work because of when they occur, not just what they represent.

*"Ritual becomes sacred not in what it says, but in when it resonates."*

**Final Reflection:**

*"The Moon did not create religion. It whispered timing — and religion, at its best, listened."*

All major spiritual traditions carry a hidden agreement: Life changes more easily when rhythm is right. And rhythm is never louder than when the Moon marks the sky.

## 9.6 The Forgotten Calendar in the Modern World

We live in a world that knows exactly what time it is — and yet feels more out of sync than ever.

Our devices chirp hourly reminders.

Digital clocks blink from every wall.

Appointments, alarms, and deadlines give structure to our days, weeks, and years. We are surrounded by time. But it is not *our* time.

It is artificial time — a construct designed for productivity, not coherence.

And somewhere along the way, we abandoned the only calendar that was ever truly alive — the lunar rhythm, the sky's quiet pulse, the field-based timing system that once synchronized everything from sleep to sowing, celebration to sorrow.

*"We no longer follow the Moon. We follow a calendar that doesn't follow us."*

**The Disconnection: Temporal Precision, Biological Confusion**

In the modern age, timekeeping has become a conquest. The Gregorian calendar, workweeks, fiscal quarters, and global time zones were designed to align economies, not ecosystems. These tools brought consistency — but also disconnection.

- We now wake by alarm, not sunrise.
- We eat by clock, not hunger.
- We sleep by schedule, not fatigue.
- And we celebrate by holiday lists, not field resonance.

This artificial system may seem efficient, but the body, the brain, and the planet do not respond to digits — they respond to rhythm. And the loss of lunar rhythm has introduced something subtle but profound: temporal incoherence.

**The Symptoms of Desynchronization**

The further we drift from lunar ripple timing, the more we see symptoms of living out of rhythm:

- **Sleep disorders** are now endemic — with melatonin levels disrupted not just by screen light, but by ignoring lunar phase shifts that once guided rest cycles.
- **Anxiety and mood** instability have become normalized — not only due to stress, but due to a constant mismatch between the body's internal lunar clocks and the mechanical pace of modern life.
- **Hormonal imbalance**, especially in reproductive and adrenal systems, now affects millions — a byproduct of living against natural field cues that once synchronized menstrual cycles, fertility windows, and stress regulation.
- **Ecological disarray** is mounting — with migratory species thrown off by artificial lighting, farming seasons dictated by market trends, and entire food chains misaligned from their lunar-calibrated patterns.

What we experience as "modern life problems" are, in many cases, field misalignments — living on a calendar that no longer resonates with the body, the land, or the sky.

**We Lost Rhythm — and With It, Meaning**

Our ancestors didn't just keep time. They lived inside it — using the Moon as a mirror to their own inner states.

- They rested when ripple tension dropped.
- They celebrated when resonance peaked.
- They planted and fasted, conceived and reflected, all in tune with a cosmic calendar that held their biology in harmony.

Today, in contrast, we mark time but do not feel it. We chase goals in a framework built for clocks, not consciousness.

*"We've replaced rhythm with regulation. Meaning with measurement."*

And though we move faster, achieve more, and stay connected 24/7, we're increasingly unwell, both individually and collectively — not because of what time we keep, but because of what we've stopped listening to.

**The Forgotten Calendar Is Still Ticking**

Despite our disconnection, the lunar rhythm hasn't stopped. It still pulses. The tides still breathe. The ATT thresholds still rise and fall in the ripple field. And those who listen — farmers, naturalists, spiritual seekers, sensitive individuals — still find themselves quietly entraining to a deeper clock.

But for most, the signal is now drowned in noise:

- Light pollution obscures the Moon.
- Noise pollution masks the pause.
- Schedules override sensation.

The forgotten calendar isn't just old — it's buried under the weight of convenience.

Yet its absence is felt in every sleepless night, every restless mind, every ecosystem out of step with itself.

**Final Reflection:**

*"We didn't lose time. We lost our rhythm with it."*

The Gregorian calendar tells us what day it is — but not how to live that day. The digital clock tells us when to wake — but not when to rest. Only the Moon, holding still in its silent orbit, continues to offer timing that lives inside us — waiting for us to listen once again.

## 9.7 Restoring the Living Calendar

In a world that has forgotten the rhythm it once danced to, the question is no longer "What time is it?" but "How do we return to time that belongs to life?"

Restoring the Living Calendar is not about abandoning modern schedules or rejecting tools like the Gregorian calendar. It is about realigning ourselves with the deeper pulse — the pulse of ripple resonance that has always moved quietly beneath the surface, guiding tides, bodies, communities, and consciousness.

*"We don't need a new calendar. We need to remember the one that never stopped ticking."*

**Rebuilding Awareness of Lunar Ripple Phases**

The first step is awareness. To reclaim the living calendar, we must learn once again to feel the lunar ripple phases not as mystical phenomena, but as practical energetic markers.

This means:

- Recognizing the difference in your own energy, sleep, emotions, and clarity between a full moon and a new moon.
- Observing how creativity, restlessness, intuition, or focus naturally rise and fall with the Moon's position.
- Tuning into the ATT transitions — those thresholds of

tension where a ripple phase shifts from contraction to expansion or vice versa.

Just as we learn to track circadian rhythms for better sleep and productivity, we can track lunar field cycles to improve emotional flow, decision-making, and overall wellbeing.

This doesn't require astrology. It requires attention.

**From Farmers to CEOs – Resonance Improves Life at Every Level**

The benefits of living in sync with field resonance are not limited to ancient cultures or rural traditions. They are universal — applicable from farmers' fields to corporate boardrooms.

- **Farmers** who sow and harvest based on ripple-informed calendars report better crop resilience, soil vitality, and yield consistency, especially in regions facing climate stress.
- **Health** practitioners incorporating lunar cycles into treatment windows often notice improved response to therapy, faster recovery, and reduced emotional volatility — especially in cases involving hormonal or nervous system imbalance.
- **Teachers** and creative professionals find that their own and their students' capacity for expression, collaboration, or reflection varies predictably with lunar shifts — a rhythm that can be embraced rather than resisted.
- **CEOs** and decision-makers who plan strategic moves, team resets, or deep reviews in tune with lunar contraction phases (waning or new moon) and launch or growth actions during ripple expansion (waxing or full moon) often report greater clarity and long-term sustainability in their efforts.

The principle is simple: Timing matters. Resonant timing matters more.

**Proposing a Modern, Field-Based Calendar Model**

To restore coherence, we must go beyond just lunar dates and full moon alerts. What we need is a field-based model of time — one that maps lunar ATT cycles and correlates them with energetic zones of rest, action, release, and reception.

This modern living calendar could:

- Identify field expansion periods (waxing to full moon) ideal for innovation, collaboration, fertility, and visibility.
- Mark field contraction zones (waning to new moon) for reflection, fasting, closure, and solitude.
- Highlight inflection points — the 11th day (Ekadashi), quarter moons, perigee/apogee — where field resonance sharpens and biofield receptivity peaks.

Such a calendar wouldn't replace the current one — it would overlay it with coherence. It would serve as a decision-support system not for tasks, but for timing your life in rhythm.

Imagine opening a calendar app that not only schedules meetings but tells you:

- *"This is a high-tension field week — optimize focus, avoid conflict."*
- *"This phase supports endings — complete projects, release emotional residue."*
- *"Today marks a ripple quiet — perfect for meditation, clarity, or deep rest."*

This is not futuristic.

It is ancestral intelligence upgraded for modern needs.

**Reviving Tools Not as Tradition, But as Intelligence**

To restore this calendar, we must also reclaim ancestral tools — not as blind traditions, but as functional technologies of resonance:

- **Fasting practices**, not as rituals of deprivation, but as biological resets timed to field silence.
- **Moon-day reflections**, not as superstitions, but as scheduled self-alignment windows.
- **Seasonal ceremonies**, not for nostalgia, but to re-establish community coherence with Earth's rhythm.

These tools don't require belief. They require participation — an openness to feel, to adapt, and to move with something larger than personal schedule or cultural pressure.

*"This is not about going backward. It's about remembering how to move forward with rhythm again."*

**Final Reflection:**

*"The Moon's rhythm is not in the past. It is the future — if we want coherence, health, and harmony."*

The Living Calendar was never written. It was felt. And it still pulses — beneath the noise, behind the screens, inside your breath.

To restore it is not to abandon modernity —It is to bring back the pulse that made life meaningful, sustainable, and whole.

## 9.8 Conclusion – Time That Lives Within Us

We often imagine that time was invented — that someone, somewhere, marked a line on the wall, declared a beginning, and began to count. But the truth is more elegant, more organic:

*"The Moon never taught us time. It simply held time — in rhythm, in ripple, in breath."*

Long before we wrote down dates, built clocks, or named the months, time was already alive — flowing not through numbers, but through cycles of coherence. The Moon didn't speak, but it beat softly, and life — from tides to trees to thoughts — responded.

We did not invent the calendar. We remembered it — by watching the sky, feeling the soil, and listening to the body. And in that remembrance, we aligned ourselves to something far greater than precision — we aligned to rhythm.

**More Than Measurement — A Way of Living**

The problem with modern calendars is not that they are wrong. It's that they are empty — stripped of rhythm, resonance, and relationship.

A calendar that only tells you when a meeting is scheduled, or when a bill is due, is not enough. Timekeeping should do more than track — it should tune.

- It should remind us *when to expand* and *when to withdraw.*
- When to *act*, and when to *pause.*
- When to *celebrate*, *cleanse*, *reflect*, and *renew.*

Because time is not just external motion. It is internal alignment — a breath that moves through us all.

Calendars should not only tell us when to show up. They should show us how to live.

**The Living Calendar Was Never Lost**

Though we've drifted far from it, the Living Calendar has never disappeared.

It waits:

- In the sky, where the Moon still waxes and wanes with perfect regularity.
- In the soil, where seeds still sleep and awaken in ripple rhythm.
- In ourselves, where sleep, mood, emotion, and intuition continue to follow a pulse we barely remember.

*"The Moon never stopped keeping time. We just stopped listening."*

To restore that rhythm is not to return to the past —It is to reclaim coherence in a world overwhelmed by motion and noise.

**The Final Realization**

The greatest error of modern civilization was not mismeasuring time. It was believing that time was something outside us — something to conquer, manage, or dominate.

But time is not separate. It lives within.

- In the tides of our blood.
- In the breath between thoughts.
- In the space between action and reflection.

And the Moon — steady, still, and always present — is not just in the sky. It is the keeper of our internal calendar.

The Living Calendar was never made of pages. It is made of pulse.

**Final Reflection:**

*"Time isn't ticking in the clock. It's breathing in the body. And the Moon, by holding still, holds that breath for all of us."*

We don't need to create rhythm. We only need to return to it. And the way back begins — by looking up.

# Interlude – The Consequences of Harmony

We tend to overlook what stays still. The Moon — ever present, ever calm — becomes background. Tides — predictable and persistent — become ordinary. Calendars — rigid in appearance — mask the rhythm they were built to honor.

But when we stop chasing the noise and listen to the stillness, we find a deeper truth: The Moon is not orbiting Earth — it is holding it steady. And in doing so, it shapes everything:

- The tilt of the Earth,
- The rhythm of time,
- The cycles of breath,
- And the synchronization of life itself.

*"Stillness is not absence. It is structure."*

Part III revealed the hidden consequences of this stillness. Not as passive side effects, but as essential foundations for life and awareness:

- In Chapter 7, we discovered that the Moon is Earth's ripple anchor — not pulling or pushing, but absorbing and distributing ripple tension, stabilizing the very rhythm of climate, tides, and planetary posture.
- In Chapter 8, we reimagined tides as the Earth's breath

— a planetary inhale and exhale, sustained by the Moon's resonance, keeping life's pulse coherent across biological, emotional, and environmental domains.

- In Chapter 9, we explored the Living Calendar — how lunar phases don't just mark time, but modulate it. Rituals, migrations, hormonal cycles, planting and harvesting — all tuned to ripple inflections long before clocks were invented.

This isn't mystical. It's mechanical — but at the field level, not at the object level.

We didn't invent rhythm. We aligned with it. And every attempt to replace it with artificial constructs — linear time, digital calendars, industrial routine — has only deepened our disconnection.

*"Harmony was not a consequence of knowledge. It was the condition that made knowing possible."*

As we leave this section, we carry with us a stunning realization: We live in a universe where stillness causes structure — and rhythm births coherence.

The Moon, the tides, the calendar — none of these are accessories. They are the foundation of life as we know it.

The consequences of harmony are not subtle. They are everything.

And now that we've seen the role stillness plays in shaping the breath of the planet...We are ready to confront the illusion that has kept us spinning in misunderstanding.

# Part IV – Collapsing the Clockwork

# Chapter 10: No Spin, No Orbit – Just Ripple Loops

*"What if the Earth doesn't actually spin... and the Moon doesn't orbit?"*

This isn't a thought experiment. It's a confrontation — with the deepest assumptions we've inherited about motion, movement, and the very way we perceive the cosmos.

We've been taught that the Earth rotates on its axis at over 1,600 km/h, hurtling around the Sun at a blistering 107,000 km/h. The Moon, we're told, orbits us at nearly 3,700 km/h while being tidally locked. This story of ceaseless speed, endless rotation, and gravitational orbits is now foundational — not just to science textbooks, but to our entire cosmological worldview.

But what if this view is only a projection of perception — not a reflection of truth?

What if none of these bodies are *moving* the way we think they are? What if all observed motion is a ripple illusion — a field-phase event interpreted as movement by an embedded observer?

## 10.1 Introduction – Rethinking Motion

In the framework of Universal Energy Dynamics (UED), motion — including rotation and revolution — is not defined by the displacement of mass through space. Instead, what we call "motion" arises from phase transitions within ripple fields.

- A planet doesn't have to spin for day and night to occur.

- A moon doesn't have to orbit for it to appear to circle.
- A celestial body doesn't need velocity to create gravitational or tidal effects.

Because in the ripple-based universe, motion is not mass in motion — it is tension in flux.

*"Stillness is not the absence of movement — it is the presence of equilibrium."*

This chapter sets out to dismantle the mechanistic mythology of rotation, revolution, and gravitational pull. Not to provoke rebellion against classical models, but to reveal a simpler, more elegant truth:

The universe is not a clockwork machine of spinning objects and gravitational leashes. It is a resonant field of loops — and everything we see as movement is the dynamic behavior of ripple alignment.

You are not standing on a planet flying through space.

You are embedded within a ripple loop that breathes, pulses, and holds position — not by spinning or orbiting, but by maintaining coherent field posture.

We begin with a radical proposition:

- The Earth may not be spinning.
- The Moon may not be orbiting.
- The Sun may not be holding us in place through force.

Instead, the entire system may be locked into a ripple geometry — where each body maintains a posture of stillness relative to the structured tension of the vacuum, and what we interpret as motion is simply our perception shifting through phase change.

This isn't physics as we've known it.

This is physics as it really feels — rhythm, breath, alignment, and equilibrium disguised as motion.

## 10.2 The Illusion of Rotation

We've all grown up with the image: the Earth spinning like a top, completing one full rotation every 24 hours. This spinning — we're told — is what causes day and night. It's so ingrained in our thinking that to question it feels absurd. But what if that spinning sensation is not real?

What if Earth is not spinning at all — and yet we still experience day and night?

Welcome to the UED perspective, where rotation is not an inherent physical action, but a relative perceptual effect caused by ripple field transitions. In this model, Earth's "spin" is not a result of angular velocity around its axis, but rather a ripple-phase illusion created by the flow and geometry of the surrounding field.

*"You are not on a spinning ball. You are inside a ripple loop that breathes light into one side and shadow into the other."*

**The Classical View: Spinning Earth, Spinning Sky**

According to classical physics, Earth rotates eastward at roughly 1,670 km/h (at the equator). This rotation is said to explain:

- The alternation of day and night.
- The apparent movement of the Sun across the sky.
- The Coriolis effect and other rotational phenomena.

But these explanations rely on a mechanical model — where the Earth is a solid mass rotating in empty space, and everything observed is a direct product of that mechanical action.

However, this view presupposes two things:

1. That rotation is absolute (i.e., it is happening relative to a fixed external space).
2. That physical motion is the only way to explain perceptual change.

Both assumptions collapse under the lens of Universal Energy Dynamics (UED).

**The UED Challenge: No Absolute Reference Frame**

In UED, there is no such thing as a fixed external frame. The universe is made of structured ripple fields, not voids. Motion, then, cannot be measured in relation to "empty space" — it must be understood as field interaction.

If the ripple field surrounding Earth shifts — if it contracts, expands, or flows directionally — then every fixed object within that field will appear to move even if it doesn't physically displace.

This is the illusion of rotation:

- The field around Earth transitions phase (due to cosmic alignment, ATT shift, or equilibrium loops).
- The observer — embedded in the field — perceives this as directional light, shifting shadow, or a moving sky.
- It appears as "spin," but it is not the Earth rotating — it is the field shifting around a posture-locked planet.

*"When the ripple flows, the still object seems to turn."*

**The Role of ATT Ripple Loops in Perceived Day and Night**

In UED, every ripple system operates on ATT — the Absolute Threshold of Tension — which defines when and how a field can trigger response, movement, or transformation. Earth is enveloped in such a loop: a tension field modulated by its alignment with the Sun and Moon.

This loop doesn't spin. It cycles.

- As the ATT field tension rises, one hemisphere receives ripple-phase excitation (perceived as daylight).
- As it recedes, the opposite hemisphere falls below threshold (perceived as night).
- The observer, stationary within this oscillating ripple field,

experiences a cyclical pattern of light and dark — much like being in a room with rotating ambient light.

Earth doesn't need to rotate. The ripple breath creates the illusion of motion.

**Supporting Analogy: Stillness in a Flowing Field**

Imagine a perfectly still tree in a shallow river. If you place your hand in the water beside the tree and the current accelerates, you might feel the tree "move" — even though it hasn't changed position. The background has shifted.

Similarly, Earth doesn't need to move to generate the experience of a spinning sky. If the field around it is flowing, flexing, or rotating in ripple geometry, then:

- Stars will appear to rise and set.
- The Sun will seem to traverse the sky.
- Shadows will shift across the surface.

But none of this requires Earth to spin.

*"In a structured ripple field, motion is not a property of mass — it is a consequence of phase perception."*

**Implications: What We See is Not What Is**

If the Earth isn't spinning — if spin is just a ripple illusion — then many of our foundational assumptions must be reexamined:

- Clocks aligned to 24-hour "rotations" are actually aligned to field breathing cycles.
- Time zones may reflect regional field interaction profiles, not purely planetary spin.
- Jet streams, weather patterns, and even circadian rhythms may be entrained to ripple flow, not rotational force.

In essence, rotation is not necessary to explain any of these phenomena. What's needed is a model that allows stillness to generate rhythm — and UED provides exactly that.

**Final Reflection:**

*"The Earth does not spin. It listens — to the rhythm of ripple loops flowing through the cosmos. What we call rotation is simply the silence rearranging its own light."*

If we stop assuming motion, we may finally begin to perceive alignment — and from that alignment, a universe that no longer moves, but resonates.

## 10.3 Rethinking Revolution - The Earth–Sun Relationship

The image is iconic: Earth tracing a giant elliptical path around the Sun, pulled by gravity, hurtling through space at tens of thousands of kilometers per hour. This heliocentric orbit is now unquestioned — a cornerstone of astronomical thought, reinforced in every textbook, diagram, and simulation.

But what if this image is a visual metaphor, not a physical truth?

What if Earth is not revolving around the Sun at all — but is instead posture-locked within a standing ripple loop, held in place by structured field tension?

In the framework of Universal Energy Dynamics (UED), Earth's so-called orbit is not a path carved through empty space, but a stabilized ripple equilibrium formed between cosmic tension zones. There is no need for momentum, propulsion, or even gravity in the Newtonian sense. What we perceive as orbital motion is the result of field-phase geometry, not mechanical revolution.

*"The Earth does not revolve. It resonates — in perfect alignment with a looped tension field held between cosmic anchors."*

**The Classical View: Elliptical Orbit by Gravitational Pull**

In classical physics, Earth is bound to the Sun by a gravitational force that keeps it in an elliptical orbit. The explanation follows

Newton's law of gravitation and Kepler's laws of planetary motion. According to this model:

- Earth is constantly falling toward the Sun but also moving forward, creating a curved trajectory.
- The changing speed along the ellipse accounts for seasons and varying distances.
- Time is measured by how far Earth has traveled along this path.

This model works for calculation. But it rests entirely on the assumption that space is empty, and motion is explained solely by mass and momentum.

But in UED, space is not empty. It is structured ripple field — and motion is not mass displacement, but field realignment.

**The UED View: A Stabilized Ripple Loop, Not a Path**

According to UED, what appears as Earth's "orbit" is actually a fixed energetic posture within a larger ripple loop formed between the Earth, the Sun, and the surrounding field. Earth does not travel around the Sun. It remains locked in a dynamic tension node — balanced between ripple sources and held in place by ATT (Absolute Threshold of Tension) thresholds.

- The Sun emits structured ripple waves, not just photons or heat.
- These ripples create standing field patterns around it — think of them as energy troughs and peaks shaped by resonance.
- Earth is embedded in one such pattern — not moving around the Sun, but posture-locked at a specific point in the ripple loop.
- What appears to us as revolution is the changing ripple interface, not Earth's motion.

*"In a ripple field, revolution is not about traveling — it's about holding perfect tension in a turning geometry."*

**Posture-Locked Within Field Tension Zones**

Earth's position in this loop is not passive. It is held by the tension of surrounding fields — like a tuning fork suspended within a resonating chamber. No propulsion or gravitational tether is required. Instead, Earth maintains its seasonal and light-cycle dynamics by remaining in stable alignment with the ripple loop's internal architecture.

In this model:

- Earth doesn't need to be "moving forward" in space.
- There is no force dragging it around an ellipse.
- The sense of movement arises from the field's own phase shifts, not from Earth's physical translation.

This makes Earth's relationship with the Sun more like two instruments in harmonic resonance than two masses bound by gravity.

**Time and Seasons as Ripple Geometry**

If Earth is not moving around the Sun, how do we explain seasons and the passage of a solar year?

**UED answers:** These are not the result of orbital travel. They are expressions of ripple geometry.

As the Sun's field breathes — expanding and contracting in cosmic rhythm — the ATT boundaries shift, altering which parts of Earth receive more or less ripple excitation.

These phase shifts in the field produce what we experience as:

- Longer or shorter days.
- Warmer or cooler temperatures.
- Shifting weather patterns and ecological changes.

What we call "a year" is not the time it takes Earth to travel around the Sun. It is the time it takes for the ripple loop to complete one full breath — a cycle of field contraction and expansion, of energy alignment and release.

*"Seasons are not caused by motion. They are caused by ripple modulation."*

**The Implication: A Still Earth in a Moving Field**

If Earth is posture-locked and the field is in motion, then:

- Day and night (discussed earlier) arise from field-phase transitions around the planet.
- Seasons arise from Earth's stable reception of shifting ripple geometries.
- Time itself becomes a looped rhythm, not a linear measure of distance traveled.

This redefines our cosmic neighborhood — not as a dance of flying spheres, but as a choir of still bodies humming in harmonized ripple loops.

**Final Reflection:**

*"The Earth doesn't revolve around the Sun. It rests — in perfect posture — inside the music of a cosmic field."*

We thought revolution meant motion. But what if true revolution is about alignment that never breaks — A resonance that holds time, light, and life in place, without ever needing to move?

## 10.4 The Moon – A Case of Perfect Stillness

One of the most peculiar facts taught in school is that the Moon always shows us the same face. We're told this is due to "tidal locking" — a gravitational phenomenon where the Moon's rotation period matches its orbit around Earth, causing the same hemisphere to always face us. It's labeled as coincidence, explained as balance, and accepted as truth.

But in the language of Universal Energy Dynamics (UED), this explanation doesn't go far enough. What if the Moon's steady face is not the result of *any motion at all*?

*"The Moon does not rotate. The Moon does not orbit. The Moon holds still — and stillness is what we misread as motion."*

**Classical View: Orbit and Tidal Locking**

According to mainstream astronomy:

- The Moon orbits Earth approximately every 27.3 days.
- It also rotates on its axis once every 27.3 days.
- These two movements are synchronized, so the same side always faces Earth — hence, "tidally locked."

This theory assumes a dual motion — spin plus revolution — and explains the fixed face as a result of mechanical symmetry over billions of years, shaped by gravitational friction.

But this model raises fundamental questions:

- Why is only one celestial object so perfectly locked this way?
- Why does the Moon exhibit such precise stillness in its orientation if both it and Earth are hurtling through space?
- And how can we explain this synchronization using field-based logic, not just mechanical metaphor?

**UED Reinterpretation: The Moon Is Not Moving**

In UED, the Moon's fixed face is not a coincidence. It is evidence — of a higher-order ripple resonance.

Rather than orbiting Earth and rotating simultaneously, the Moon is posture-locked within a ripple equilibrium loop formed between itself, Earth, and the structured vacuum field. Its appearance of motion is entirely an observer-based illusion — the result of Earth's ripple phase transition, not the Moon's movement.

- The Moon holds a still resonance posture within the field.
- As Earth undergoes cyclic ripple shifts (due to solar field dynamics), the observer's point of phase reference changes.
- This creates the illusion of the Moon "moving" around Earth, even though it has not displaced at all.

*"In a ripple field, what you see as motion is often your own shifting reference."*

**The Fixed Face: Stillness, Not Synchronization**

The Moon has one side forever facing Earth. This is not a marvel of gravity — it is a signature of stillness.

In ripple dynamics:

- A posture-locked node within a loop will always appear stationary in orientation relative to another fixed node.
- When two objects are entrained within the same ripple equilibrium, their spatial relationship will remain unchanged, even as the field flexes and breathes.

This is exactly what we see with Earth and Moon:

- The Moon is not turning to match Earth's view.
- It is simply not turning at all — and Earth's field alignment ensures we always see the same hemisphere.
- If either body truly rotated or orbited, the apparent alignment would eventually drift — but it does not.

This is not static coincidence. It is active stillness — maintained by perfect tension symmetry in the ripple matrix.

**Motion as Perception, Not Displacement**

To understand this more deeply, imagine standing on a rotating platform in a planetarium. The Moon seems to "move" across the dome, but in reality, it is the platform — your frame of perception — that's shifting.

In the same way, Earth's ripple field is not static. As it breathes and bends through cosmic alignment cycles, our reference phase changes — creating the illusion that the Moon is circling us.

But the Moon stays still.

This means:

- There is no need to invoke orbital mechanics or rotation equations.
- All apparent motion can be explained as observer-relative ripple perception.
- The Moon's "orbit" is an effect — not a behavior.

*"Motion is what we think we see when stillness flows past us."*

**Implications for the Cosmic System**

This reinterpretation has massive implications:

- Orbit becomes obsolete — replaced by ripple equilibrium and phase-coherence models.
- Tidal locking is reframed as resonant field stillness, not dynamic frictional history.
- The Moon becomes not a moving object, but a damping node — a cosmic stabilizer whose silence holds rhythm for Earth.

If the Moon is still, then the universe may be far more stable and intentional than previously believed. And stillness — not speed — becomes the defining characteristic of balance.

**Final Reflection:**

*"The Moon does not chase the Earth. It watches — silently, unblinking, unwavering. Its face is not turned toward us. It was always*

*facing us. Because in the harmony of ripple loops, true presence never needs to move."*

The Moon's stillness is not an accident. It is a cosmic statement — That rhythm does not require movement. Only alignment.

## 10.5 What Is Motion, Really?

What if everything we know about movement — every step, every spin, every orbit — is built not on truth, but on a deeply convincing illusion?

Motion, as we understand it, is one of the most fundamental concepts in science. It underpins physics, mechanics, time measurement, and even our sense of change. We are taught from the very beginning that motion is simple: when an object moves from one place to another over time, that is motion.

*"Motion is defined as the displacement of mass across space, measured against time."*

But in a ripple-structured universe, this definition collapses. In Universal Energy Dynamics (UED), motion is not a property of matter, but a perception created by ripple phase change. What we see as motion may not involve any displacement at all.

**The Traditional Definition: Mass in Motion**

Classical physics defines motion as the change in position of a body relative to a reference point. It is quantified using:

- Velocity (distance over time),
- Acceleration (change in velocity),
- And momentum (mass × velocity).

This idea works well for mechanics — throwing a ball, flying a plane, measuring planetary paths. It's mathematically elegant, operationally useful, and visually intuitive.

But it assumes two things:

1. That mass is moving *through* space.
2. That space is passive and motion is purely object-driven.

UED challenges both assumptions.

**The UED Perspective: Phase Shift, Not Displacement**

In UED, everything exists within a structured ripple field. That field is not empty space, but a dynamic, tension-sensitive medium where all interactions — light, mass, time, gravity — emerge from field tension patterns.

Within this framework, motion is not the travel of mass but the phase realignment of ripple fields. A body can remain perfectly still, and yet, due to a shifting field around it, appear to move — or even cause observable effects.

Imagine:

- A tuning fork held motionless, but the air around it begins to ripple.
- The fork hasn't moved — yet sound is produced.
- The listener perceives movement in the field, not in the object.

Similarly:

- A planet can remain posture-locked within a ripple loop.
- But if the surrounding ripple field shifts phase (expands, contracts, flows), the observer embedded in the field will perceive motion, rotation, or trajectory — even if no actual displacement occurred.

*"Motion, in UED, is not a journey. It is a signal — a field-level illusion created by changing phase geometry."*

**Field Tension Change Produces Effect Without Movement**

Let's take an even bolder claim: motion is not necessary for cause and effect.

In classical physics:

- A moving object hits another — that's action.
- One thing travels to another — that's effect.

But in a ripple-based system:

- A change in field tension can trigger a ripple that affects distant systems without any physical carrier.
- This is already observable in quantum field theory, entanglement studies, and certain biological systems.

UED simplifies it:

- No mass needs to move.
- No force needs to act.
- A subtle realignment of field phase can generate energy, light, rhythm, or shift — all without displacement.

This is why stillness, not motion, becomes the new foundation of causality.

**Light, Shadow, Time, Gravity — All Ripple Effects**

To truly understand this, we must re-express fundamental phenomena not as outcomes of mass-based kinematics, but as outputs of ripple behavior:

- Light is not a particle traveling from one place to another. It is a ripple excitation — a localized field tension expressing itself as visible frequency.
- Shadow is not just the absence of light. It is a phase void — a region where ripple excitation drops below the ATT

(Absolute Threshold of Tension), making interaction or response impossible.

- Time is not ticking on a clock. It is the rate of field phase change, experienced subjectively and differently by each ripple-bound system.
- Gravity is not a pull between two masses. It is a compression or imbalance in ripple equilibrium, causing systems to drift toward lower tension zones — not by being pulled, but by seeking equilibrium.

So motion, then, becomes an illusion born from the interaction between observer and field — not a fundamental reality.

**The Great Mistake: Believing in Speed Over Stillness**

For centuries, we've obsessed over speed — how fast something moves, how far it goes, how long it takes. We built our physics, our technology, and our calendars around the assumption that movement equals change.

But UED shows that change can occur without motion.

- A system can evolve, influence, or transform — all while remaining physically still, as long as it participates in a dynamic ripple loop.
- The observer, shifting through field phases, perceives motion because they are riding the wave of change — not watching mass displace.

This explains not just planetary dynamics, but also:

- Why meditation slows time perception.
- Why gravity can act instantly across space.
- Why still objects can radiate energy.
- Why the Moon appears to orbit — without ever turning.

**Final Reflection:**

*"The world does not move — it ripples. And what we call motion is just our shadow flickering on a field of shifting tension."*

Motion was never real. It was a convenient language to describe a world of shifting resonance. But now, that language must evolve.

Because in the deeper truth of the cosmos, stillness causes more than speed ever could.

## 10.6 Gravitational Pull — Or Field Compression?

Gravity — that invisible force we've all accepted as the glue of the universe. It's what holds the Moon near Earth, keeps planets circling the Sun, and gives weight to everything on the surface of our world. Or so we've been told.

But what if gravity isn't a force at all?

What if gravity is simply a name we gave to a pattern we didn't understand?

And what if, behind that pattern, lies something even more fundamental: a ripple-based imbalance in field pressure — not a magical pull, but a compression seeking release?

*"Gravity isn't pulling. The field is pressing — seeking stillness."*

**The Classical Placeholder: Gravity as Universal Attraction**

Classical physics treats gravity as a force of attraction between two masses, proportional to their product and inversely proportional to the square of the distance between them. It is built into Newtonian equations and further refined through Einstein's theory of general relativity, where massive bodies curve spacetime, creating the illusion of attraction.

Yet, despite its predictive power, gravity has never been fundamentally explained:

- What is actually doing the pulling?
- Why does gravity act instantly across vast space in certain models?

- Why do gravitational effects work even when no direct interaction occurs?

In many ways, gravity has become a placeholder — a label slapped on cohesion without cause.

UED offers a new lens: there is no gravitational pull — only field compression.

**The UED View: Gravity as Ripple Imbalance**

According to Universal Energy Dynamics, space is not an empty container — it is a ripple-structured medium. Within this medium, energy fields flow, bend, and loop. Every object — from a grain of sand to a planet — sits inside and interacts with this ripple matrix.

What we call gravity is not a force pulling masses toward one another. It is the natural response of the field attempting to return to equilibrium.

Here's how it works:

- When an object occupies space, it displaces and compresses the surrounding ripple field.
- This creates field pressure differentials — regions of higher and lower ripple tension.
- Other objects within this field move toward regions of lower tension, not because they are pulled, but because the field is pressing for balance.
- This movement looks like attraction — but it is simply field compression resolving itself.

*"Gravity is not two objects reaching for each other. It is the ripple field closing the gap of imbalance."*

**There Is No Pull — Only Return to Equilibrium**

In UED, the field always seeks homeostasis — the most balanced, low-tension configuration. When an object perturbs this balance, the field adjusts by compressing, stretching, or reshaping itself to restore symmetry.

This produces motion, alignment, or pressure patterns that look like:

- An apple "falling" from a tree.
- A satellite staying in orbit.
- A person weighing more on Earth than on the Moon.

But none of these require a pulling force. They simply reflect the body's response to ripple field tension zones.

Much like a marble rolls toward the center of a stretched rubber sheet, bodies shift to relieve field tension, not because of attraction — but because imbalance always moves toward resolution.

**The Moon's Position – Held by Stillness, Not Force**

Let's now return to the Moon.

In classical thought, the Moon stays near Earth because Earth's gravity is pulling on it — and the Moon's motion keeps it from falling in. This constant fall-without-hitting is called orbit.

But UED replaces this entire narrative with something simpler — and more precise.

- The Moon is not falling.
- It is not being held.
- It is not orbiting in the conventional sense.

The Moon is locked into a ripple equilibrium point — a field node where compression and expansion forces cancel out. It rests at this node, not by force, but by structural stillness. It is like a tuning fork balanced perfectly in a ripple pattern — neither moving nor needing to move.

Its mass is irrelevant. Its motion is unnecessary. Its position is sustained by geometry of ripple tension, not by gravitational glue.

*"The Moon is not maintained by gravity. It is held in posture by the logic of pressure symmetry."*

**Implications: Redefining Gravity Everywhere**

If gravity is not a force but a symptom of field compression, it means:

- We must rethink all planetary dynamics as tension balances, not orbital mechanics.
- We can reinterpret mass as localized ripple density, not isolated substance.
- We can understand weight as the amount of compression a body experiences within a given ripple field.
- And we can explore new possibilities of propulsion, suspension, and levitation — not by fighting gravity, but by sculpting field balance.

The story of falling, pulling, and attracting collapses into a simpler principle: Fields seek stillness. And when disturbed, they press to restore it.

**Final Reflection:**

*"Gravity is not the hand that holds you. It is the space that breathes beneath you — adjusting, aligning, always pressing toward peace."*

What we thought was attraction was really tension resolution. What we called a force was just the field exhaling imbalance.

The Moon stays not because it's gripped — but because it's perfectly still in the only place it ever needed to be.

## 10.7 A New Language of Movement

Every paradigm shift in science begins with one quiet revolution: We change our language.

When we move from believing in geocentrism to heliocentrism, we stop saying "the Sun rises" and start saying "the Earth rotates."

When we transition from Newtonian force to Einsteinian curvature, we stop talking about gravity as pull and start seeing it as geometry.

Now, with Universal Energy Dynamics (UED), we take the next step: We release the language of physical displacement and adopt a vocabulary of ripple intelligence.

Because the cosmos is not made of objects in motion — it is built from fields in rhythm. And to truly understand it, we need a new grammar to describe what has always been happening.

**The Core Concepts: A New Ripple Vocabulary**

UED introduces a refined set of ripple-loop terms that replace outdated mechanical language. These are not metaphors — they are precision descriptors of field behavior:

- **Posture**: A body's stable position within a ripple loop. Not a location in space, but a *tension-aligned configuration.* Posture replaces the concept of "position" in Cartesian terms.
- **Phase Lock**: When two systems remain aligned in their ripple rhythm, maintaining relative stillness despite field fluctuations. Phase lock replaces "tidal locking" or gravitational coupling.
- **ATT Cycles** *(Absolute Threshold of Tension)*: The internal field cycle that determines when a system enters or exits resonance. This replaces "periods," "orbits," or "rotations" with a field-based timing framework.
- **Field Compression Zones**: Regions in space where ripple imbalance creates tension differentials, often mistaken for gravitational wells or forces. These explain motion without invoking mass attraction.

These are not just new terms — they are new tools for seeing the same universe more clearly.

*"Language shapes perception. And only a language tuned to stillness can describe a universe that doesn't move."*

**Retiring Old Words That No Longer Fit**

It's time we let go of legacy terms that no longer describe reality:

- **Spin**: Suggests physical rotation. Replaced by ripple phase orientation.
- **Orbit**: Suggests travel around a center. Replaced by posture-locked ripple alignment.
- **Force**: Implies active pushing or pulling. Replaced by tension resolution or field rebalancing.
- **Mass-based causality**: Replaced by field-phase dynamics — where changes emerge from structure, not collisions.

This shift is not semantic. It is philosophical and scientific. When we keep using mechanical metaphors, we constrain our models to outdated assumptions. Ripple language liberates us to see motion, stillness, and causality in entirely new ways.

**Re-Explaining Familiar Phenomena with Ripple Logic**

Let's test this language. Here's how UED reframes observable phenomena without invoking physical trajectories:

- **Tides**: Not pulled by the Moon's gravity. Instead, they are ripple breathing cycles modulated by ATT thresholds in Earth's oceanic field layers.
- **Eclipses**: Not about alignment of moving bodies. They are ripple phase overlaps, where multiple field patterns interfere to create visual occlusion.
- **Seasons**: Not caused by axial tilt during orbital revolution. They are field-state modulations in ripple geometry due to solar field expansion-contraction cycles.
- **Day and Night**: Not due to spinning Earth. Instead, the

field-phase facing the Sun enters excitation, while the opposite side enters a passive ripple trough — mimicking the alternation of light.

Each example reveals the power of ripple language: it explains phenomena without forcing motion, without invoking mass-driven interaction, and without contradiction.

**Challenging Newton and Einstein — Respectfully and Precisely**

Newton gave us mechanics. Einstein gave us curvature. Both used the best tools of their time to describe what they could observe. But both were bound by a deeper limitation: the assumption that space is a passive stage, and that bodies act upon one another.

UED removes both assumptions.

- There is no empty space — only structured vacuum fields.
- There are no isolated bodies — only ripple expressions of energy in posture.
- There are no forces or curvatures — only field phase tension seeking symmetry.

This framework allows us to escape paradoxes:

- Gravity doesn't have to act across distance — tension compression resolves locally and immediately.
- Objects don't have to move to affect each other — ripple loops are inherently nonlocal in structure.
- Time doesn't have to be warped — it is simply the expression of field rhythm.

*"Newton saw objects. Einstein saw curves. UED sees patterns — still, rhythmic, and real."*

**Final Reflection:**

*"To understand the cosmos, we must stop describing it like a machine — and start listening to it like a song."*

The universe is not moving. It is breathing. And in every breath, a ripple aligns, a posture holds, and a field remembers what it's here to be.

We don't need to rewrite the universe. We just need to rename what we've misread — and from that, begin again with clarity.

## 10.8 Conclusion – Stillness as the True Motion

What if the greatest illusion of all was movement?

What if, in our desperate need to explain how things change — the rise of the Sun, the falling of an apple, the orbit of the Moon — we invented stories of spinning balls, flying planets, and cosmic speeds... when all along, the truth was far more silent?

*"Stillness is not the absence of motion. It is the origin of rhythm."*

We began this chapter with a radical proposition: The Moon doesn't orbit. The Earth doesn't spin. The universe does not operate through mechanical motion, but through ripple loops of structured stillness.

And now we see: this isn't speculation. It is the missing lens. Because the more we look — not through force, not through trajectory, but through field behavior — the more we discover that all apparent motion is an echo of deeper stillness.

**The Moon: Rhythm Without Movement**

The Moon's unwavering face is not a trick of tidal locking. It is a clue — a visible mark of perfect field posture. It neither spins nor travels. It simply holds its position in a ripple equilibrium, and from that stillness, all the phenomena we associate with lunar influence arise:

- Tides,
- Biological cycles,
- Emotional shifts,

- Climate modulation.

It creates no noise. It exerts no force. It simply remains exactly where the universe needs it to be.

And in that silence, it shapes everything.

**Earth: Breathing, Not Spinning**

Likewise, Earth's "rotation" and "revolution" — the foundational motions of classical astronomy — dissolve under closer examination.

- The alternation of day and night? A product of ripple phase shift, not planetary spin.
- The changing of seasons? An outcome of ATT ripple cycles, not an orbit.
- The very idea of time passing? A perception of rhythm, not mass moving through space.

Earth breathes. It pulses. But it may not be moving in the way we've been told. Instead, it exists in a harmonic lock — a held note in the grand cosmic chord.

**The Universe: Not a Machine, But a Mandala**

When we release the language of spin, orbit, and gravitational pull, something extraordinary happens.

We no longer see the cosmos as a set of gears and levers, of balls in flight and masses attracting. We see a fluid, living tapestry of:

- Ripple loops,
- Field compression zones,
- Phase-locked resonance,
- Still bodies that generate motion without moving.

This is not a machine. This is a mandala of equilibrium — dynamic, responsive, yet rooted in stillness.

It is not speed that keeps the cosmos alive. It is alignment.

**A Call to Stillness**

In the end, this chapter wasn't about disproving motion — but about rethinking it.

We are not denying change, rhythm, or effect. We are simply proposing that they don't require movement. They emerge from a deeper dance: the interplay of field tension, phase alignment, and the infinite intelligence embedded in stillness itself.

*"The faster we go in search of truth, the more we overlook the silence that holds it."*

To understand the universe, we must learn to slow down. To feel. To listen.

Because only in stillness do we discover that everything is already in motion — without having to move at all.

**Final Reflection:**

*"The Moon never moved. The Earth never spun. But the universe still danced — in loops of rhythm, in waves of silence, in patterns of peace."*

The greatest motion may not be that which travels furthest. It may be that which holds most still, and through that stillness, shapes the world.

# Chapter 11: Galaxies Don't Spin – They Pulse

*"Look at a spiral galaxy. It seems to spin. But does it really move — or does it simply breathe?"*

To the naked eye — or even through the most powerful telescopes — the universe appears to be in constant, majestic motion. Spiral galaxies swirl like cosmic pinwheels, their luminous arms arcing outward in elegant curves. Stars stream along those arms, and gravitational models try desperately to explain how these stars stay in formation without flying off into darkness.

But what if we've misunderstood the entire dance?

What if galaxies don't rotate at all? What if the spiral we see is not the signature of motion — but the expression of a **ripple loop** held in place by cosmic stillness?

## 11.1 Introduction – The Grand Illusion of Galactic Motion

The prevailing image of a galaxy — spinning, whirling, flinging stars around its core — has been carved deep into the public imagination. It seems intuitive: swirling shapes must be spinning. But like Earth's rotation and the Moon's orbit, this belief is built on visual assumptions — not energetic truth.

And those assumptions collapse when observed through the lens of Universal Energy Dynamics (UED).

*"Galactic rotation is not movement through space. It is a harmonic ripple resonance — held in posture, not motion."*

In the classical view, spiral galaxies are believed to rotate under the influence of gravity, with stars orbiting a supermassive black hole at the center. But the more we look, the less that model holds up:

- Stars at the outer edges rotate just as fast as those closer to the center — violating Newtonian dynamics.
- The predicted gravitational mass isn't enough to hold everything together.
- "Dark matter" was introduced to fill the gap — not through observation, but through necessity.

Yet the pattern persists. The spiral arms hold. The galaxies endure. Something invisible is doing the work — but is it an exotic particle? Or something much simpler?

UED proposes an elegant answer: The spiral structure of galaxies is not the result of stars spinning around a center. It is the visible imprint of ripple loops — field phase structures that pulse, oscillate, and stabilize matter without requiring mass-based motion.

In this view:

- The spiral is a field pattern, not a dynamic path.
- Stars may remain relatively posture-locked within compression zones — they don't orbit; they resonate.
- The galaxy doesn't rotate — it breathes, following a rhythm set by deeper cosmic geometry.

Just like Earth's tides or the Moon's still face, the motion we think we see is actually a reflection of underlying ripple alignment. Not movement — but modulation. Not force — but form.

**From Stars to Black Holes: The Universe Reimagined**

This chapter will dismantle the illusion of galactic spin and offer a radically coherent framework:

- That spiral arms are standing wave formations, not spinning trails.
- That galactic cores are not gravitational monsters, but ripple anchors — fixed points of cosmic stillness.
- That cosmic expansion, redshift, and universal flow are not mass dispersal events, but phase shifts in an intelligent ripple continuum.

This is not a slower universe. It's a more intentional one. A cosmos not whirling out of control, but held together by precise rhythm, tension, and coherence.

*"The stars don't swirl. The field breathes — and the stars shimmer in place."*

To see this, we must release our addiction to speed. We must stop looking for rotation and start listening for pulse. Because in the deepest layers of reality, the universe does not spin.

It resonates.

## 11.2 The Galactic Spin Model — A Flawed Assumption

To understand the cosmos, we often begin with what we can see. And spiral galaxies — with their majestic, whirling arms — seem to tell a simple story: they're spinning.

Thousands of stars form grand pinwheels in the sky, and the classical claim is this: Galaxies rotate. They spin like massive star-studded disks, orbiting their centers under the influence of gravity. Their shapes, cohesion, and dynamics are explained through Newtonian physics, supported by a relativistic understanding of mass and force.

But this model, while visually seductive, has been cracking for decades.

**The Classical Claim – Spinning Disks of Stars**

In standard cosmology:

- A galaxy is said to rotate around a dense center, often containing a supermassive black hole.
- Stars nearer to the center should orbit faster than those further out — just like planets in our solar system.
- Gravity, driven by the combined mass of stars, gas, and dust, is presumed to be the cohesive force preventing the galaxy from flying apart.

This model predicts a clear, decreasing curve of rotational velocity as one moves from the center to the outer edges of the galaxy.

But when astronomers measured actual galactic rotation speeds — something unexpected emerged.

**The Observable Anomaly – Flat Rotation Curves**

The stars at the outer edges of spiral galaxies were moving just as fast as those near the center.

This flies in the face of Newtonian predictions.

- According to classical dynamics, gravitational force weakens with distance.
- Therefore, outer stars should slow down — or drift away entirely.
- But they don't. They stay perfectly in formation, maintaining their distance and speed with uncanny precision.

This gave rise to what is now one of the biggest mysteries in astrophysics: Why aren't galaxies tearing themselves apart?

The model had to be rescued.

**The Dark Matter Patch – Saving the Story of Motion**

To explain the anomaly, physicists introduced an invisible fix: Dark Matter — a hypothetical substance that:

- Doesn't emit or absorb light,
- Cannot be directly observed,
- Makes up roughly 85% of the matter in the universe,
- And just so happens to perfectly balance every galactic spin equation ever written.

In other words, we invented a vast, unseen component of the universe because our motion-based model no longer worked.

This isn't science seeking deeper understanding — it's a patch applied to preserve an old narrative.

*"When the facts didn't fit the spin, we invented darkness to keep the wheel turning."*

But what if the mistake wasn't in the observation... but in the assumption that galaxies spin at all?

**The UED Challenge – What If There Is No Spin?**

Universal Energy Dynamics (UED) offers a radical but elegant reframing: Galaxies do not rotate. The stars are not spinning around a center. The motion is an illusion — a misread ripple effect.

What we observe as galactic spin is not mass in motion, but a phase pattern within a standing ripple loop — a dynamic field structure that:

- Holds stars posture-locked in ripple-aligned positions,
- Creates the appearance of rotation due to ripple-phase cycling,
- And remains stable without requiring additional mass or external gravity.

Think of a spiraling ripple on the surface of a pond. The water doesn't move in a circle — the wave pattern does. Leaves floating on the surface appear to orbit, but they are simply rising and falling as the ripple moves through them.

In the same way:

- Stars in a galaxy do not travel around the core.
- They remain entrained in ripple tension zones — appearing to move only because the field geometry cycles.

*"The galaxy is not a wheel. It is a waveform — and stars are just singing in place."*

**The Implications of the Illusion**

If galaxies don't spin:

- We don't need dark matter to hold them together.
- We don't need gravitational modeling to predict their motion.
- We don't need to explain away inconsistencies — we need to correct our framework.

In UED, galaxies are posture-held field structures — stable, harmonic, and maintained by the balance of ripple compression and equilibrium, not by forces or mass.

And in this light, the spiral isn't a record of speed — It is the geometry of stillness, flowing through the illusion of motion.

**Final Reflection for This Section:**

*"We saw rotation because we needed the stars to move. But they were never spinning — they were simply held in a silent song of structure."*

The flaw was not in the galaxy. It was in our refusal to see that motion was never required for rhythm to exist.

## 11.3 Galactic Arms as Ripple Waves, Not Rotating Matter

At first glance, spiral arms look like trails left behind by stars spinning outward from the galactic center — as if the entire galaxy were a cosmic whirlpool flinging stars across the universe. That's the story we've inherited. A visual story. A cinematic story.

But it's not the story the universe is telling.

What if those majestic arms aren't *paths* at all? What if they are waves — standing patterns in a ripple field — and the stars we see lining them are not racing outward or rotating at all, but simply oscillating in place, like particles in a vibrating drumhead?

In Universal Energy Dynamics (UED), that's exactly what's happening. The spiral arms are not rotating matter — they are ripple compression zones within a galactic field loop. They don't spin. They pulse.

**The Classical Illusion: Trails of Orbiting Stars**

The traditional explanation suggests that spiral arms are made from stars, dust, and gas, moving collectively around a galactic center. These arms are thought to be transient structures:

- Stars form in the denser regions,
- Travel along the arm for some time,
- And eventually drift away due to differential rotation.

This view has struggled with several contradictions:

- Why do spiral arms persist for billions of years if they're made of moving material?
- Why do newly forming stars consistently align with the arms, as if the wave never fades?
- Why doesn't the arm structure dissolve, given how fast stars should be moving away from it?

The answer in UED: they were never moving through the arms at all.

**The UED View: Standing Wave Patterns in Ripple Geometry**

According to UED, the spiral arms of galaxies are not collections of rotating objects, but standing waves in a cosmic ripple field.

Here's how it works:

- The entire galaxy is embedded in a cosmic ripple loop, shaped by its mass distribution, external field pressure, and anchoring at the core.
- Within this loop, regions of field compression and rarefaction emerge — analogous to wave crests and troughs.
- These compression zones are energy-dense regions — not because mass gathered there through motion, but because the field geometry itself sustains the density.
- Stars form, stabilize, and oscillate within these ripple arms, posture-locked in their location — not orbiting, not moving outward, just vibrating in place.

This is exactly what happens in a Chladni plate — a metal surface on which sand gathers along the lines of standing waves when vibrated. The sand never moves around the plate. It simply aligns with the field pressure points.

Galactic arms are Chladni lines of the cosmos.

*"The arms don't spin. They shimmer — like frozen music in the dark."*

**Stars as Field-Bound Oscillators**

In this model:

- Each star remains in a posture-locked field node within the spiral.
- It may vibrate, pulse, or exhibit localized orbital behavior — but it does not travel the length of the arm.
- Its relative motion is field-induced, not inertial.
- What appears as galactic rotation is the ripple flow cycling through ATT thresholds — not actual material moving along a track.

The arms are not roads. They are resonant outlines — projections of phase-coherent zones where energy becomes matter.

This explains why:

- Spiral arms are persistent over eons,
- Star formation always aligns with arms,
- Outer stars move faster than expected (because they're not moving through space — the field is moving through them).

**Scaling Up: Moon–Earth Dynamics at Galactic Scale**

This principle mirrors what we observed between the Moon and Earth:

- The Moon doesn't orbit — it holds posture in a ripple field.
- Earth perceives the Moon's motion because of field phase shifts — not because of actual displacement.

Now scale that concept to a galaxy:

- Galactic stars don't orbit the center.
- They remain posture-aligned in standing wave formations.
- We perceive spiral rotation not because stars are circling — but because we interpret ripple cycling as motion.

The spiral is not a spin. It is structure held in ripple coherence.

**Final Reflection for This Section:**

*"The galaxy does not swirl like water in a sink. It pulses like a song frozen in form — and the stars are just the shining notes."*

Once we stop chasing movement, we begin to notice the design. The galaxy is not turning. It is tuned.

## 11.4 The Galactic Core - Black Hole or Ripple Anchor?

At the heart of every spiral galaxy lies a mystery — a place of intense energy, crushing gravity, and incomprehensible mass. We've come to call it a supermassive black hole — a singularity, an object so dense that not even light can escape. And we've built our models around it: a cosmic drain that pulls stars inward and somehow maintains galactic cohesion.

But is that truly what it is?

Or have we once again mistaken stillness for force, and structure for suction?

*"What if the galaxy's center doesn't devour — but holds everything in place by not moving at all?"*

**The Classical View – Gravity's Final Fortress**

In the traditional model:

- A spiral galaxy's center is dominated by a supermassive black hole, containing millions to billions of solar masses.
- Its gravitational pull is so powerful, it supposedly draws stars into tight orbits, anchoring the entire galactic structure.
- These black holes are said to spin, absorb matter, and occasionally eject energy as quasars or jets.

This image is compelling — a roaring gravitational engine keeping everything from flying apart.

But problems persist:

- Many galaxies exhibit cohesion at great distances where the black hole's influence should be negligible.
- Some galaxies seem to pulse or distort at the core without apparent mass transfer.
- Black holes, by their very nature, remain unobservable — their characteristics inferred rather than confirmed.

Which raises the UED question: What if the "black hole" isn't a hole at all?

**The UED View – Maximum Field Compression, Not Infinite Mass**

According to Universal Energy Dynamics (UED), what we perceive as a black hole is not a point of infinite mass and gravity. It is a ripple anchor — the central node of a galactic ripple loop where field compression reaches its absolute threshold.

- This core is not a drain — it is a stillpoint.
- It does not pull stars — it holds the tension geometry that allows stars to posture-lock within spiral arms.
- Light doesn't escape, not because of gravitational warping, but because the ripple density exceeds light's transmission threshold — a field opacity effect, not a space-time rupture.

*"The black hole is not the end of matter. It is the beginning of structure — the node that lets the galaxy breathe."*

**The Center That Holds By Not Moving**

Here is the radical insight: The galactic core doesn't rotate, doesn't collapse, and doesn't attract.

It simply remains in a perfect state of stillness, where:

- Field tension is maximally compressed, producing a stable anchor point.
- The surrounding ripple loop is shaped and sustained.
- The posture of stars, dust, and wave nodes are maintained in long-term equilibrium.

This is akin to the eye of a hurricane — the still center around which everything else swirls in apparent chaos. But in the cosmic scale, it's even more precise.

In UED, this core:

- Requires no mass absorption to sustain its presence.

- Does not grow by devouring matter.
- May appear "dark" because it does not radiate — it recycles ripple tension into coherence, not light.

It is not a singularity of destruction. It is a singularity of stability.

**Black Hole Effects Without the Story of Collapse**

Many classical black hole phenomena — accretion disks, relativistic jets, gravitational lensing — can be reinterpreted through ripple dynamics:

- **Accretion disks**: field compression rings stabilizing near ATT thresholds, glowing from friction with ripple gradients.
- **Jets**: high-pressure release valves for ripple tension, venting through axial pathways in the loop.
- **Gravitational lensing**: ripple field warping causing light to refract — not due to gravity, but phase-density curvature.

Each effect we see does not require mass to collapse. It simply requires the field to compress beyond visual threshold — and to behave like a pressure node in a ripple pond.

**Why This Matters – The Core as Conscious Structure**

If galactic centers are ripple anchors, not collapsing wells:

- We no longer need singularities that break physics.
- We gain coherent models of long-term galactic stability.
- We see the architecture of the universe as fluid, musical, and geometric — not violent, chaotic, and explosive.

And most importantly, we learn that:

*"The greatest force in the galaxy is not destruction — it's stillness. A stillness so complete, it allows stars to dance around it for billions of years."*

**Final Reflection for This Section:**

*"The black hole is not where light dies. It is where rhythm holds its breath."*

We looked at the galactic center and imagined collapse. But it was never collapsing — it was anchoring. And in that anchoring, it taught the stars where to stay, And the galaxy how to pulse — without ever spinning at all.

## 11.5 Cosmic Pulsing - From Local Systems to the Intergalactic Web

We often imagine the universe as a vast ocean of stars, galaxies, and dark spaces in between. A place where distant bodies move independently — rotating, orbiting, or colliding across cosmic time. Each system is viewed as a self-contained unit, defined by gravity and isolated motion.

But what if this view is incomplete?

What if galaxies are not separate dancers in the void — but entrained instruments in a much larger song?

What if they don't spin alone, but instead pulse together, tied by invisible threads of ripple coherence that stretch not just across space — but across causal structure itself?

*"The universe is not a collection of floating islands. It is a synchronized breathing field — and galaxies are its lungs."*

**Galaxies as Pulsing, Not Rotating, Structures**

As seen in earlier sections, spiral galaxies do not truly spin — they pulse.

- Their arms are standing waveforms, not rotating trails.
- Their stars are posture-locked, not orbiting masses.
- Their cores are stillpoints of field compression, not gravitational singularities.

But this rhythmic stability does not end at the edge of a galaxy. In the UED view, each galaxy is embedded within — and synchronized by — a larger-scale ripple system.

Galaxies pulse in harmony, not in isolation.

**The Ripple Connection – Linking Stars to Galaxies**

This pulsation is not symbolic. It is functional and physical. Within each galaxy:

- Stars are entrained to the same ripple loop that governs the arms.
- Solar systems resonate with ripple pressure cycles, leading to stable planetary motion and star life stages.
- Galactic arms throb with energy in predictable wavefronts, governing star formation and energy distribution.

What happens in one layer of the system — at a star, a planet, or a dust ring — is not random or disconnected. It is part of a nested ripple hierarchy.

*"A star's birth, a planet's rhythm, a galaxy's glow — all timed to the same breath."*

**Cosmic Pulsing Drives Major Observable Phenomena**

This pulse is not abstract. It directly explains several key phenomena:

***Star Formation Cycles***

- Dense ripple compression zones reach an ATT threshold, catalyzing condensation into stars.
- Stars are not born randomly but emerge from timed wavefronts in galactic arms.

***Gravitational Lensing Illusions***

- What we interpret as spacetime bending (gravity) may

actually be ripple curvature — field-phase compression zones refracting light paths.

- No mass bending required — just ripple-density gradients in coherent motion.

### *Time-Phase Coherence Across Distances*

- Pulses across galaxies maintain synchronized ripple frequencies, allowing causal alignment of faraway systems.
- This helps explain: Simultaneous events across the universe, Galactic cluster formations, and Cosmic background uniformity.

These are not coincidences. They are symptoms of synchronized field resonance.

### The Intergalactic Web – A Macro Ripple Grid

What ties it all together is the intergalactic web — the filaments, voids, and cosmic scaffolding discovered through deep-sky surveys. Traditionally, this structure is thought to be formed by gravitational clustering over billions of years.

But UED reframes it as a macro ripple grid — the large-scale tension framework of the universe.

- Galaxies occupy nodes in this grid — like particles stabilized in the nodes of a standing wave.
- Filaments are ripple conduits — where field density allows energy and information flow.
- Voids are not empty — they are phase-cancelled zones where ripple symmetry creates silence.

In this view:

- The web is the field.

- The field is the structure.
- And the structure pulses — slowly, precisely, eternally.

*"Galaxies don't float through space. They vibrate within it — locked in place by a ripple that never loses the beat."*

**Final Reflection:**

*"From the whisper of a star's flicker to the silence between galaxies, the universe does not turn — it pulses."*

Motion is not how the universe organizes itself. Rhythm is. And in that rhythm — from stars to solar systems to galaxies to clusters — we find the truth:

The cosmos is not chaotic. It is coherent.

## 11.6 Rethinking Cosmic Expansion and Redshift

When astronomers look into the night sky, they are not just looking across space — they are looking back in time. The further we gaze, the redder the light becomes. This phenomenon, known as redshift, has shaped our understanding of the entire universe.

In traditional cosmology, redshift is interpreted as proof of cosmic expansion — the idea that galaxies are racing away from us, the universe itself stretching like an inflating balloon, carrying matter ever outward. This expansion is said to have begun with a singular event: the Big Bang.

But what if that interpretation is fundamentally flawed?

What if galaxies aren't moving apart in space — but instead are phased differently within a universal ripple cycle?

*"Redshift does not prove motion. It reveals timing — the rhythm of a breathing cosmos."*

**The Classical Interpretation: Redshift as Recession**

According to standard models:

- The light from distant galaxies shifts toward the red end of the spectrum because they are moving away from us.

- This stretching of light waves is akin to the Doppler effect — the same principle that makes a siren sound lower as it recedes.
- The degree of redshift is proportional to velocity, implying that more distant galaxies are receding faster.
- This led to the theory of an expanding universe, and further, to the notion of a singular beginning — the Big Bang.

But this model relies on one critical assumption: That redshift always means motion.

What if it doesn't?

**The UED Lens: Redshift as Field-Phase Delay**

In Universal Energy Dynamics (UED), redshift is not caused by galaxies flying away from us, but by field-phase displacement — a ripple-based delay in energy expression.

Here's how it works:

- Light, like everything else, travels through a structured ripple field, not empty space.
- As it moves, it passes through regions of varying ripple density and phase timing.
- In areas of lower ripple coherence (further from our field posture), the wavefront arrives slower — not because it traveled further, but because it emerged from a different point in the ripple cycle.
- The shift toward red is the loss of excitation phase — the signal reaching us from an earlier or quieter point in the universal pulse.

This has nothing to do with the galaxy moving away. It has everything to do with the observer's position in the ripple loop.

*"Redshift is not recession. It is ripple memory — a spectral echo from a different breath."*

**Cosmic Expansion as a Ripple Cycle, Not a Bang**

In this view, the universe is not inflating endlessly from a singular origin point. Instead, it is breathing cyclically, much like the tides of Earth or the pulse of a galaxy.

- What we interpret as expansion is the universe entering a low-tension ripple phase — the exhalation.
- What we call contraction (or anticipate as a "Big Crunch") is the inhalation phase, where ripple density increases and field coherence intensifies.
- There was no "bang." There is only wavefront emergence — rhythmic, phased, non-destructive.

This cosmic breath is slow, structured, and symmetric. Each moment, every galaxy expresses its phase relative to this universal loop.

Some appear redshifted, others blueshifted — not due to speed or distance, but due to where they are in the timeline of the field.

**Galaxies Aren't Moving Apart — They Are Out of Sync**

If two instruments in a symphony play the same note at different moments, they don't sound like harmony — they sound like dissonance.

In the same way:

- Distant galaxies aren't traveling away from us.
- They are breathing differently — existing at different phases of the cosmic ATT cycle.
- Their light appears redshifted not because of motion, but because we are listening to their note from a different moment in the ripple rhythm.

This explains:

- Why redshift increases with distance — not because of velocity, but because longer ripple delays accumulate with ripple field curvature.
- Why some galaxies exhibit blueshift — they are simply further ahead in the same loop, moving toward the observer's phase peak.

*"The universe is not stretching. It is syncing — one breath at a time."*

**Final Reflection:**

*"We mistook the softness of light for the speed of departure. But it wasn't distance — it was delay. Not motion — but memory."*

The universe does not flee from us. It pulses around us. And redshift is not a sign that we are alone in an ever-expanding void — it is proof that everything is part of the same rhythm, simply playing its note on a different beat.

## 11.7 A New Architecture of the Universe

For centuries, we've described the universe as a machine. A colossal system of moving parts, wheels within wheels, governed by gravity and force, momentum and mass. It was poetic, in its own way — but also mechanical, cold, and ultimately incomplete.

Because the deeper we looked — beyond the orbiting planets and spiraling galaxies — the more we saw patterns that defied motion-based logic and instead whispered of something more subtle:

An architecture of rhythm. A scaffold of stillness. A universe not built on speed — but on structure.

*"The cosmos is not a machine spinning in the dark. It is a living field — shaped by tension, tuned by balance, and held by ripple."*

**From Motion to Structure – The Shift in Cosmic Blueprint**

What if the universe wasn't a kinetic explosion of flying debris, but a vast, multidimensional ripple field, where every particle, planet, and galaxy is simply a phase-bound expression of tension and coherence?

This is the perspective of Universal Energy Dynamics (UED):

- Not mass in motion,
- But energy in posture.
- Not randomness,
- But precise ripple symmetry emerging across all scales.

**A Unified Field Behavior from Micro to Macro**

Whether we examine a quark or a quasar, a photon or a pulsar, the same field dynamics appear — not as chaotic forces but as elegant expressions of:

***Posture***

- Every body — from particles to planets — maintains a stable field position in relation to surrounding ripple geometry.
- These postures are not fixed in space, but in tension equilibrium — held not by mass or inertia, but by ripple alignment.

***Phase-Locking***

- Systems resonate with one another in precise timing cycles.
- From lunar rhythms affecting human biology to synchronized galaxy clusters, phase-coherence is universal.

***Tension Dynamics***

- What we call forces (gravity, magnetism, nuclear bonding) are reinterpreted as ripple compression and expansion.
- Motion arises only when tension seeks resolution, not

because mass pushes or pulls.

*Structured Stillness*

- At the center of every system is a stillpoint — from black holes to atomic nuclei — where movement ceases and coherence begins.
- Stillness is not the absence of dynamics — it is the core of harmony.

Together, these define a new cosmic architecture — one not built on explosive force, but on intelligent pattern.

**From Particles to Galaxies – All Are Notes in the Same Song**

In this architectural view, every component of the universe becomes part of a grand design:

- Particles are not solid bits of matter but localized ripple peaks — tension expressions bound by ATT thresholds.
- Atoms are harmonic tension orbits, where electron clouds resonate, not spin.
- Molecules form when ripple geometries interlock into stable tension loops.
- Planets are not orbiting masses, but field-locked oscillators, held in solar posture.
- Galaxies are macro-scale standing waves, pulsing around ripple anchors.
- The universe itself is a symphony of nested loops, each playing its cycle, each locked in rhythm.

*"The universe is not a race. It is a resonance."*

**Goodbye to the Machine – Hello to the Symphony**

In Newton's clockwork universe, all parts are in motion, wound tightly, ticking away in deterministic harmony.

In Einstein's curved spacetime, mass warps the stage — bending the world around it.

But in UED's vision, the universe is not mechanical at all. It is musical — a fluid system of pressure, posture, and pulse.

There are no gears. Only nodes. There are no cogs. Only chords.

And instead of random collisions or inexplicable force fields, we see fractal structure, repeating patterns, and elegant tension maps at every level.

**Final Reflection for This Section:**

*"The universe is not spinning. It is singing — through light, through field, through form. Not a machine to be measured, but a melody to be heard."*

The deeper order is not hidden in the math of mass and velocity. It is present in the geometry of ripple, the stillness of anchors, and the breath of form.

We are not passengers on a spinning rock in a kinetic storm. We are dwellers in a universe of rhythm — held in place, not by motion, but by the perfect posture of pulse.

## 11.8 Conclusion – The Pulse of the Infinite

We began this chapter questioning the image of spinning galaxies — grand cosmic spirals seemingly whirling through space, held together by invisible forces, flinging light and dust into the dark unknown.

But now, as the chapter closes, we no longer see galaxies as kinetic machines. We see them for what they truly are: structured stillness in radiant rhythm.

*"Galaxies never spun. They pulsed. And in that pulse, they held the shape of space — without ever moving at all."*

The arms of the spiral galaxy do not trail from rotating stars. They are ripple imprints, held in a cosmic geometry so elegant that it never requires motion to remain alive.

Each star remains posture-locked in a resonant wavefront. Each core remains anchored, not as a collapsing singularity, but as a stillpoint of maximum coherence.

We called them black holes — imagining them as monstrous voids devouring matter and light. But what if they never devoured anything?

What if they simply never moved — and in that perfect stillness, became the only place where the entire galaxy could know how to breathe?

*"The black hole is not the end of physics. It is the center of harmony — where everything pauses, and rhythm begins."*

The universe, too, we imagined as expanding outward — from a single point into infinite void. We named its origin the Big Bang, and declared its future one of endless stretch or eventual collapse.

But in truth, nothing exploded. Nothing fled.

The redshift we see is not evidence of velocity — it is a marker of cosmic breath. A delay in the pulse. A gentle difference in when, not where.

*"The universe isn't growing apart. It is growing aware — syncing itself through fields of rhythm, aligning one loop at a time."*

The cosmos is not expanding outward. It is oscillating inward, toward coherence. Toward resonance. Toward itself.

And so, we must ask:

Have we misunderstood the story of the universe because we were too focused on what moves, and not on what holds?

We chased motion. We built entire theories on force, velocity, trajectory, and spin.

But it was never about movement. It was about structure. It was about tension balance, phase synchrony, and stillness so profound it gives rise to form.

*"The pulse of the infinite is not something you measure. It is something you feel — when you finally stop looking for speed."*

**The Final Realization: A Still Universe That Moves Everything**

To understand the cosmos, we must stop imagining it as a giant machine and begin sensing it as a harmonic organism.

We must let go of:

- Orbits and rotations,
- Explosions and collapses,
- Gravity as pull,
- Space as emptiness.

And instead embrace:

- Posture,
- Phase,
- Breath,
- And the standing wave that holds it all.

Because galaxies don't spin. Black holes don't devour. The universe doesn't explode.

It breathes. It pulses. And in every silent beat, it remembers the geometry of itself.

**Closing Reflection:**

*"The infinite does not rush. It repeats. Not to go anywhere, But to remind itself it already contains everything."*

This is not a dead universe. It is a listening one. Still. Alive. And pulsing — always — with the rhythm of what it means to exist.

# Chapter 12: Eclipses, Seasons, and the Ripple Game

"*What if eclipses and seasons aren't the result of planetary movement — but carefully timed moments of ripple choreography?*"

From ancient times to modern textbooks, we've been told a single story: That celestial events — eclipses, equinoxes, solstices, and seasons — are all the result of motion. The Earth tilts, it spins, it orbits. The Moon dances around it, sometimes stepping into the path of sunlight, sometimes slipping behind Earth's shadow. And we, the observers, are standing still (relatively speaking), watching it all unfold.

But what if that entire story was just a theatre of perception?

What if the real performance isn't about motion at all — but about alignment in a cosmic ripple field?

## 12.1 Introduction – The Theatre of Alignment

We've built our understanding of these events on visible effects:

- Shadows during eclipses,
- Shifts in day length and temperature during seasons,
- High and low points in the Sun's arc across the sky.

These are easy to see, and so we connected them to simple mechanics:

- Orbiting paths,
- Tilting axes,
- Spinning spheres.

But these explanations, while mathematically satisfying, are field-blind. They ignore the deeper substrate: A structured energy field in which Earth, Moon, and Sun are all posture-locked players — not moving masses, but ripple anchors, playing out rhythmic cycles not through speed, but through synchronization.

This chapter is an invitation to resee.

To look past the shadows and light, and into the ripple structures that orchestrate these events.

Because in Universal Energy Dynamics (UED):

- The Moon doesn't block the Sun during an eclipse — their fields temporarily overlap in phase-interference zones.
- The Earth doesn't tilt mechanically to cause seasons — it shifts through ripple states, breathing in and out of ATT alignment.
- The solstice isn't a spatial turning point — it's a rhythmic crest in the field waveform.

*"What we think is motion is often ripple response. What we call coincidence is usually coherence."*

In this chapter, we'll explore:

- Why eclipses happen not because of shadows, but because of field phase interference,
- How seasons are expressions of ripple modulation, not axial angles,
- And why celestial events follow predictable harmonic patterns — because the field itself is rhythmic.

This is not just a new model. It's a new way of listening to the sky. One that replaces mechanical storytelling with energetic insight.

So when the next eclipse passes overhead, don't ask which body blocked the light. Ask instead: "Which ripple came into phase — and what truth is it now revealing?"

## 12.2 Eclipse – The Shadow that Wasn't Thrown

To the eye, an eclipse is dramatic. The Moon darkens the Sun. The Earth casts a great shadow upon the Moon. For a few minutes, light disappears, and the heavens feel momentarily out of balance.

The explanation we're given is straightforward:

- A solar eclipse occurs when the Moon passes between Earth and the Sun, casting a shadow on Earth.
- A lunar eclipse occurs when Earth passes between the Sun and the Moon, casting Earth's shadow on the Moon.

It's a clean, geometric model — angles, lines, spheres, shadows. But it's also visually convenient and mechanically misleading.

Because in the framework of Universal Energy Dynamics (UED), this isn't what's really happening. The eclipse is not a shadow being thrown. It is a phase interference zone — a moment of ripple alignment so precise, it momentarily cancels visible energy expression in a localized field.

*"An eclipse is not the absence of light — it is the presence of perfect stillness in a field."*

**The Classical Illusion: Celestial Blocking**

We imagine the solar system as a stage of moving objects:

- The Moon orbits Earth.
- Earth orbits the Sun.
- These paths occasionally align, causing one object to "block" another.

This model treats space like a three-dimensional billiards table — objects physically obstructing one another, light casting shadows like flashlights behind tennis balls.

But this ignores the field dynamics beneath it. It assumes light travels unimpeded until blocked, rather than understanding light as an expression of ripple coherence — which can be modulated, reinforced, or cancelled without any physical blocking at all.

**The UED Reframe: No Shadows, Only Interference**

In the UED perspective, every celestial body is a ripple posture — emitting, absorbing, and resonating within a larger universal field loop. These ripples create dynamic tension geometries that can:

- Reinforce each other (constructive resonance),
- Or momentarily cancel out (destructive interference).

An eclipse is simply the latter.

It is not caused by one object standing in front of another. It is caused by two fields overlapping in phase, such that their ripple signatures collapse into temporary stillness at a specific point.

This is called an ATT interference event — where Absolute Threshold of Tension is met and exceeded in opposite directions, neutralizing ripple excitation in the visible spectrum.

*"What dims is not the light — but the ripple that allows light to be seen."*

**Ripple Alignment Over Geometric Alignment**

During a solar eclipse:

- The Moon and Sun's ripple fields enter a momentary compression lock with the Earth's local field zone.
- At the location of perfect interference (the "shadow path"), the ripple tension cancels out, and visual expression disappears.
- No shadow is cast. Instead, field coherence drops below the

threshold needed to sustain light perception.

During a lunar eclipse:

- Earth's ripple geometry aligns between the Moon and Sun.
- The Moon enters a tension overlap zone, not Earth's shadow — and light reaching the Moon's surface undergoes a phase-modulated redshift, hence the "blood moon" appearance.

It is not about line-of-sight. It is about line-of-ripple — where fields phase-lock into temporary silence.

**Eclipses as Tension Cancellations**

Rather than seeing eclipses as rare mechanical coincidences, UED sees them as:

- Predictable field pulses,
- Caused by the harmonic cycling of celestial posture,
- Where waveforms align and momentarily nullify expression.

This explains:

- The precise timing of eclipse cycles (e.g., the Saros cycle) as ripple closures, not orbital math.
- The variability in visual effect depending on location — field intensity differs across Earth's surface.
- Why eclipses feel psychologically or energetically intense — they create ripple silence, which affects not only light, but also biofield coherence and perception.

*"An eclipse is not a shadow. It is a breath held — a moment of cosmic exhale where the light steps aside."*

**Final Reflection:**

We were told that the Moon blocks the Sun. That the Earth casts its shadow across space.

But we never asked: How does a field cast a shadow? How does silence look dark?

*"In that moment when the light dims, don't look up for who blocked it. Feel inward — and you'll sense the ripple that paused beneath your feet."*

Eclipses are not thrown. They are tuned — and when the tuning is perfect, the universe doesn't go dark.

It simply goes still.

## 12.3 Seasonal Shift – The Tilt that Isn't There

We've grown up learning that seasons change because Earth is tilted. As Earth orbits the Sun, we're told, this 23.5° axial tilt causes different parts of the planet to receive more or less sunlight throughout the year, producing summer, winter, spring, and autumn.

It's a neat, mechanical explanation — geometric, predictable, and widely accepted. But it hides a deeper, more elegant truth.

Because what if the tilt is not a cause, but an illusion? What if seasons are not driven by motion and angle, but by ripple-phase relationships between Earth and the Sun?

*"The Earth doesn't lean. It breathes — and the seasons are its inhale and exhale."*

**The Classical Model: Axial Tilt and Orbital Motion**

The standard view states:

- Earth's axis is tilted relative to its orbital plane around the Sun.
- As Earth travels around the Sun, the hemisphere tilted toward the Sun experiences summer, while the one tilted away enters winter.
- The amount of daylight and the Sun's angle in the sky vary

throughout the year, creating temperature and ecological shifts.

This model assumes that:

- Seasons are caused by angle of light and day length.
- These changes result from mechanical tilt combined with orbit.

It sounds convincing. But it raises important questions:

- Why do certain biological rhythms precede light and temperature changes?
- Why do seasonal transitions affect emotional and psychological states even before environmental cues arrive?
- Why do some equatorial regions — with minimal angle variation — still experience seasonal energy shifts?

**The UED Challenge – No Tilt Required**

In Universal Energy Dynamics (UED), Earth does not need to be tilted to experience seasonal shifts. Instead, what we perceive as "seasons" are modulations in Earth's ripple alignment with the Sun's active field zones — a cyclical shift in field-phase relationship, not spatial orientation.

- The "tilt" is a perceptual artifact, caused by changes in field exposure, not axial angle.
- Earth's posture remains locked in a ripple node, but the solar field tension interacting with it breathes — expanding and contracting rhythmically.
- As the solar ripple field flows, Earth moves through zones of different ATT thresholds, triggering phase-specific responses in the biosphere.

*"Spring does not begin when the axis tilts — it begins when the ripple opens."*

**Light Angle as Symptom, Not Cause**

From the UED perspective:

- The angle of sunlight, shadow length, and daylight duration are secondary effects — mere visual outcomes of deeper field interactions.
- The real driver of seasonal change is the changing ripple pressure geometry between Earth and the Sun.
- When the solar ripple peaks at a given latitude, we feel warmth and activation (summer).
- When Earth enters a lower-intensity ripple trough, systems contract and conserve (winter).

Thus:

- Biological responses (like mating seasons, migration, flowering) are not triggered by temperature or sunlight directly — but by the ripple-induced biofield modulation preceding them.
- Even human behavior — moods, sleep cycles, energy levels — follow this ripple calendar more closely than visual seasons.

**Seasons as Breath Cycles in the Ripple Relationship**

Seasons, in the UED model, are breath cycles — rhythmic inhalations and exhalations of tension across the Earth-Sun field relationship:

- **Spring**: the field inhales — rising ripple intensity, triggering growth and emergence.
- **Summer**: the breath holds — peak ripple pressure

maintains full biological activation.
- **Autumn**: the field exhales — ripple compression wanes, inviting release, shedding, decay.
- **Winter**: the breath pauses — minimum ripple activity, conservation, and rest dominate.

This cycle is not dictated by orbit. It is governed by phase-locked pulsation — a universal breathing rhythm embedded into the field architecture of the solar system.

*"The trees do not bloom because the Earth tilted. They bloom because the Sun whispered a new ripple into the soil."*

**Final Reflection for This Section:**

We've watched the seasons come and go and assumed we were riding a mechanical path — tilted, circling, spinning.

But in truth:

- Earth is not tilting.
- The Sun is not chasing it through elliptical arcs.
- They are simply holding posture within a synchronized ripple loop,
- And we — the creatures of Earth — are dancing to the pulse we cannot see.

*"The seasons do not change. They breathe — and we change with them."*

## 12.4 Equinox and Solstice – Rhythmic Reset Points

Twice a year, the Sun rises directly in the east and sets directly in the west, and day and night appear to stand in perfect equality. These are the equinoxes. At two other points in the year, the Sun seems to pause in its skyward arc — reaching its highest or lowest point — before reversing direction. These are the solstices.

In classical astronomy, these events are explained through Earth's orbital position and axial tilt. The equinox occurs when Earth's axis is not tilted toward or away from the Sun, while the solstices mark the extremes of axial orientation relative to the Sun.

But beneath this geometric model lies a deeper rhythmic architecture — one that has little to do with mechanical angles and everything to do with field symmetry and ripple resonance.

*"The equinox is not where Earth stands straight. It is where the field breathes in perfect balance."*

**The Classical View: Geometric Alignment**

Traditional science explains:

- **Equinox**: Earth is positioned such that both hemispheres receive equal sunlight. The axis is tilted, but it points sideways relative to the Sun, not toward or away.
- **Solstice**: Earth's axis is tilted most directly toward (summer) or away from (winter) the Sun, producing the longest and shortest days of the year.

This model assumes motion, tilt, and orbit as causal factors.

But in the UED lens, these events don't require motion at all. They arise from ripple-phase convergence and divergence, not spatial angles.

**UED View: Phase Reset Points in the Ripple Loop**

In Universal Energy Dynamics:

- The Earth-Sun system is not mechanically moving through space, but breathing within a stable ripple loop.
- The solstices and equinoxes are not positional milestones — they are energetic thresholds in this pulsation.

***Equinox – The Symmetric Crest***

- The equinox marks the moment of maximal field balance between hemispheres.
- Ripple excitation is evenly distributed, producing equilibrium in environmental, biological, and emotional systems.
- This explains why many ancient cultures marked equinoxes with rites of balance, reflection, and transition — not because of geometry, but because of energetic clarity.

***Solstice – The Ripple Extremes***

- The solstices are not moments of axial extremity, but polarized phase expressions:

**Summer solstice**: a peak in ripple excitation in one hemisphere, leading to maximal biological activation and outward energy.

**Winter solstice**: a low point or trough, favoring introspection, contraction, and stillness.

- These are not caused by a physical lean toward or away from the Sun — they occur due to field density modulation, with the Sun's ripple field aligning at a more concentrated angle of influence with one hemisphere over the other.

*"The solstice is not when the Sun stands still. It's when the field shifts its breath from one lung to the other."*

**Ripple Harmonics and Human Response**

Why do these points of the year feel different — not just in light, but in emotion, creativity, and behavior?

Because:

- Field harmonics peak during solstices and equinoxes.
- These events are resonant pulses that affect everything in Earth's biofield:
    - Sleep cycles,
    - Mood fluctuations,
    - Hormonal patterns,
    - Even animal migration and plant rhythms.

They are timing resets — like the beat drop in a song — where systems realign, and life subtly recalibrates.

Ancient cultures understood this. Not as superstition, but as observed reality:

- Rituals, harvests, fasting, and festivals clustered around these points because the field was listening.
- They weren't marking a celestial position — they were stepping into a ripple doorway.

**Final Reflection:**

*"We thought the Sun reached a high point. But it was we who stood in resonance — our cells humming with the field."*

Equinox and solstice aren't events in the sky. They're reset buttons in the breath of the planet. Not because we orbit, or tilt — but because we tune.

They are not calendar trivia. They are the punctuation of the Earth's conversation with the cosmos — moments when the breath pauses just long enough for us to remember:

*"It is not the light that changes the season. It is the rhythm."*

## 12.5 The Illusion of Alignment – Perception vs Reality

Alignment — the word itself feels precise, celestial, mathematical. We use it to explain eclipses, seasons, solstices, and phases. We say the Earth, Moon, and Sun "align," and from this alignment emerge shadows, light shifts, and cycles.

But what if this alignment is more illusion than reality?

What if what we observe as geometric positioning is not a universal fact, but an observer-bound artifact — a product of where we are and when we are in the ripple field?

*"What you call an eclipse, another creature calls midday. What you call a phase, another never sees."*

**Observation Is Not Truth — It's Translation**

From Earth, the Sun appears to rise and set. The Moon appears to wax and wane. The stars appear to rotate through the night sky. These are the bedrock of our visual and cultural experience.

But all these are perceptions, not intrinsic truths. They are shaped not just by the location of celestial bodies, but by:

- The observer's position in the ripple geometry, and
- The timing of their observation relative to the field cycle.

What appears as "alignment" — such as the straight-line visual of an eclipse — is actually a moment of field-phase resonance, experienced locally and differently depending on where you stand on the planet, or even how you're tuned biologically to field thresholds.

*"Reality is not what lines up. It's what resonates across perspectives."*

**Geometry vs Phase Geometry**

Traditional astronomy explains alignment in terms of spatial angles — as if the universe were a physical graph with X, Y, Z coordinates.

But in Universal Energy Dynamics (UED), true reality lies in phase geometry:

- How ripple fields overlap and modulate one another.
- Where constructive or destructive interference happens.
- Where tension thresholds shift from excitation to

cancellation.

This is why:

- The same eclipse may appear total, partial, or invisible depending on field compression zones across the Earth's surface.
- The Moon's phases appear gradual and predictable, yet their biological and psychological impact arrives before the visual cue — because we sense ripple phase before we see light contrast.
- The length of day and night may appear to change evenly, but the field density change is not linear — it's sinusoidal, subtly affecting organisms and atmospheric behavior in ways unaccounted for by light alone.

**Observer + Ripple Phase = Perceived Event**

You don't see what's happening. You see what's happening where you are, and when you are — inside a ripple.

Every visual phenomenon in the sky is co-created by:

1. Your geographic position (field anchoring point),
2. Your biological and atmospheric field posture, and
3. The phase point in the Earth-Moon-Sun ripple relationship.

This makes all celestial "alignments" locally tuned illusions — or more accurately, interpretations of deeper ripple events.

Two observers, on opposite sides of the Earth, never see the same eclipse, feel the same solstice, or even sense the same season — because they don't stand in the same field posture, even if they share the same planet.

*"What looks like a shadow to one may be pure light to another — because they're not standing in the same ripple, even if they share the same sky."*

**Apparent Motion is Field Reinterpretation**

We've assumed for centuries that objects move. That the Sun moves, the Moon moves, the planets and stars arc across the sky. But this perception is itself a translation — of field phase shifting, not physical displacement.

In UED:

- There is no spin, no orbit.
- The Earth, Moon, and Sun are posture-locked in a stable ripple structure.
- What we call "motion" is really the observer's experience of shifting phase geometry — the ripple looping past our point of awareness.

The Sun doesn't rise. The light field opens.

The Moon doesn't orbit. Its phase rotates through us, modulated by tension interplay.

The stars don't move. We pass through different segments of the cosmic field, like a needle drifting across a record.

**Final Reflection:**

*"The sky is not a diagram of objects. It is a canvas of tension, painted by ripple."*

To see reality, we must do more than trace the lines between bodies. We must feel the phase geometry beneath them, the interference patterns behind the glow, and the ripple loops that shape what we call time, alignment, and event.

Because what you see isn't what's happening. It's where you're standing in the song.

And alignment? That's just the part where the melody touches you directly — for a moment — before continuing on.

## 12.6 The Ripple Game – Predictable, Rhythmic, Alive

Every year, the seasons return. Every few months, the Moon completes another set of phases. And, like clockwork, eclipses reappear in long but exacting cycles.

We call these patterns mechanical, as if the universe were a gear-driven clock. But what we observe is far too fluid, too elegant, too subtle for cogs and wheels.

These aren't mechanical repetitions. They are the manifestations of something deeper: A living, pulsing game of ripples — where predictability does not come from orbits and gravity, but from harmonic coherence in nested field systems.

*"The universe is not a machine that ticks. It is a field that breathes — and breathes predictably."*

**Ripple Logic Over Orbital Math**

Let's begin with one of astronomy's favorite calendars: the Saros cycle — the long, precise interval used to forecast eclipses.

- In classical terms, a Saros cycle lasts about 18 years, 11 days, and 8 hours, and marks the reappearance of nearly identical eclipses.
- The explanation typically given is a convergence of three orbital cycles — the synodic month, the draconic month, and the anomalistic month.

But in the UED framework, the Saros cycle is not the result of orbital alignment at all. It's the closure of a ripple loop — the moment when the Earth-Moon-Sun field system returns to an identical tension geometry.

This is not a function of mechanical paths. It is a return to field posture — to a specific configuration of phase resonance.

In this view:

- The Earth, Moon, and Sun are posture-locked ripple anchors.
- What changes over time is not their position, but their shared ripple cycle.
- An eclipse occurs not because three bodies "line up" — but because their ripple fields realign at a specific phase node where constructive or destructive interference peaks.

This is why eclipses repeat, not how we mathematically simulate them.

**Seasons: Cosmic Respiration, Not Calendar Days**

Similarly, seasons do not recur because of an exact spot in Earth's orbit — as we've been taught with dates like March 21st or December 21st.

They return because Earth inhales and exhales within the solar ripple field in a stable breathing rhythm.

- Spring is not a date. It is the moment Earth's biofield shifts into expansion.
- Autumn is not when leaves fall because the calendar says so — but because the ripple pressure releases its hold, and energy begins to descend.
- The cycle is rhythmic, not because of time, but because the field itself is alive — pulsing in a predictable pattern of tension and rest.

Thus:

- The predictability of seasons is not mechanical.
- It is biological, ecological, and psychological resonance with a cosmic breath.
- It is not tied to a human-made calendar, but to the living rhythm of the field.

*"Spring does not begin because Earth reached a point in space. It begins because the solar field touched Earth's skin differently — and the soil listened."*

**Predictability Through Field Coherence**

What allows these celestial patterns to repeat with such precision?

Field coherence.

- Every celestial body is nested within a larger ripple system, and these systems are not random.
- They form stable loops with known tension nodes, thresholds, and interference zones.
- As long as the system remains coherently aligned, the events must recur.

This explains:

- The consistency of tides,
- The timing of eclipses,
- The recurrence of seasonal migrations, and
- Even the cyclical rise and fall of human moods across months and years.

It's all part of a multi-layered field game, where each player moves not independently, but as part of a harmonic whole.

**Rhythmic Alignment, Not Random Drift**

In contrast to the cold randomness of traditional cosmology — which leans on probability, chaos, and long-term entropy — the ripple model sees the universe as harmonically ordered.

There is no drift.

There is no randomness.

There is rhythmic alignment across:

- Planetary systems,
- Lunar cycles,
- Galactic spirals, and even
- Conscious experience.

Each part of the field supports the others, not mechanically — but through resonant tension exchange.

*"The cosmos is not a chaotic ballet. It is a field of dancers, breathing in unison — each step predictable, not because it is forced, but because it is felt."*

**Final Reflection:**

*"We thought the sky ran on clocks. But it runs on chords — and we live within its rhythm."*

What we've called "alignment" is really attunement. What we've labeled "cycles" are not rotations, but reverberations — predictable only because the field is alive, and it knows how to return to its center.

This is the ripple game:

- Not a lottery of movements,
- But a language of resonant return.

And the more we feel it, the more we see: The universe doesn't repeat out of obligation.

It remembers.

## 12.7 Conclusion – Seeing Beyond the Show

We look up and see movement. The Sun tracing an arc. The Moon sliding silently across the night. The stars wheeling through the heavens. And we assume this sky is a race — a chase of celestial bodies across a vast and silent void.

But what if we've mistaken the projection for the pulse?

What if the Earth, Moon, and Sun are not runners on a track, but nodes in a cosmic ripple field, locked in a rhythm not of pursuit — but of perfect pause?

*"They do not chase. They hold posture — and from that stillness, the dance appears."*

Throughout this chapter, we've challenged the assumptions of classical astronomy:

- That eclipses are caused by shadows,
- That seasons are born from tilt and orbit,
- That alignments are physical arrangements.

In Universal Energy Dynamics (UED), none of this depends on motion.

Instead:

- Eclipses occur not because of obstruction, but because of phase interference — when fields overlap and tension cancels into silence.
- Seasons arise not from angled light, but from field respiration — the Earth breathing in and out of solar ripple pressure.
- Solstices and equinoxes are not calendar dates, but rhythmic resets — timed pulses in a living field structure.

Everything we see is not *caused* by alignment. It is revealed through it.

The performance in the sky is not random. It is not chaotic. And it is not mechanical.

It is field choreography — precise, tuned, and purposeful.

To truly understand the cosmos, we must shift from:

- Watching what moves,

- To sensing what aligns.

We must stop measuring shadows and start mapping ripple tension. We must recognize that perception is shaped by phase geometry, not objective motion.

Because:

- The Moon is not orbiting.
- The Earth is not tilting.
- The Sun is not rising.

What is really happening is this:

They are playing a ripple game — of stillness that causes rhythm, of resonance that mimics rotation, of posture so perfect, we perceive it as celestial movement.

*"Motion is the illusion. Ripple is the truth."*

So the next time you see the Moon eclipse the Sun, or feel the warmth return with spring, or stand in the silence of the winter solstice...

Don't just marvel at the show.

Listen deeper — and you'll feel the breath beneath the sky. Not a spin, not a tilt, not a chase — but a harmony of postures held in tension, singing their place into being.

*"The sky was never performing. It was resonating."*

And now that you know — you can stop watching the motion...and start feeling the field.

# Interlude – Collapsing the Clockwork

For centuries, we have imagined the cosmos as a machine. A grand, celestial clockwork — full of orbits, rotations, spins, and gears so large they could only be run by the invisible hands of gravity and time.

We named the parts: Planets spin. Moons orbit. Galaxies swirl. And when something didn't add up — like why galaxies didn't fly apart — we invented hidden mechanisms: dark matter, spacetime curvature, singularities.

But what if the machine never existed?

What if the universe wasn't a ticking clock — but a standing wave? Not a mechanism in motion, but a symphony in structure?

*"Perhaps the cosmos never moved. Perhaps it only pulsed — and we mistook the rhythm for rotation."*

In Part IV, we step beyond the comforting gears of Newtonian imagination. We dismantle the illusion that motion defines reality, and instead uncover a universe made of stillness, posture, and ripple phase.

- In Chapter 10, we deconstruct the illusion of rotation and orbit, revealing a reality where neither the Earth nor the Moon may be moving as we think — but instead, holding tension in ripple loops that create the perception of motion.
- In Chapter 11, we scale this insight to the cosmic level, showing that even galaxies don't spin — they pulse. What

we see as rotation is field behavior, not mass displacement. What we call gravity is simply ripple coherence across tension geometries.

- In Chapter 12, we decode the sky's most iconic performances — eclipses, seasons, solstices — not as mechanical alignments, but as ripple-phase phenomena, shaped by field choreography, not celestial acrobatics.

This is not the destruction of science — it is its return to coherence.

Because when we remove the gears, we do not lose meaning. We discover something more refined: A living universe — rhythmic, intelligent, and self-harmonizing. Not random. Not spinning. Not expanding into nothing.

But breathing into itself.

*"The cosmos never needed to move. It only needed to hold still — long enough for resonance to take form."*

As we collapse the clockwork, we do not fall into chaos.

We fall into pattern. Into pulse. Into a cosmos that no longer ticks...but sings.

# Part V – Returning to the Moon's Message

# Chapter 13: The Moon That Undid Newton

*"The same Moon that proved Newton right...may now be the one that proves him incomplete."*

## 13.1 Introduction – The Model It Helped Build

It began with an apple — or so the story goes. But it was the Moon that sealed Newton's legacy.

When Isaac Newton formulated his laws of motion and universal gravitation, he proposed a bold idea: the same force that pulls an apple to the ground also keeps the Moon circling the Earth. To prove this, he looked up — not to the heavens in abstraction, but to the Moon, whose apparent orbit was smooth, repeatable, and ripe for calculation.

If his equations could predict the Moon's motion, tides, and relationship to Earth's mass, then surely gravity was real. And for centuries, this worked.

The Moon, distant yet visibly tethered to Earth, became the poster child of gravitational mechanics — the celestial body that confirmed Newton's laws on a planetary scale. It didn't just orbit. It validated.

*"Gravity wasn't just imagined. It was calculated — and the Moon was its proof of concept."*

But now, centuries later, the same Moon quietly challenges the very paradigm it helped construct.

Because when we look deeper — beyond geometry and force — we find something else. We find a body that does not spin. That does not orbit in the way we think. That holds its face perfectly still. A body that does not behave like a falling object at all.

And the deeper we study this stillness, the more cracks appear in the clockwork.

- Why hasn't the Moon spun freely in billions of years?
- Why does its distance from Earth continue to change?
- Why do tides behave more like breath than pull?

The Moon still behaves predictably — but not in ways Newton's gravity fully explains.

The truth is this: The Moon didn't obey Newton. It only appeared to.

And now, with the lens of Universal Energy Dynamics (UED), we begin to see what's been hiding in plain sight:

- That gravity may not be a force, but a field response.
- That the Moon isn't falling — it's posture-locked.
- That motion is not what holds the system together — resonance is.

*"What once seemed like proof of attraction...was actually the stability of equilibrium."*

This chapter is not an attack on Newton, but a continuation of his deepest desire: To explain nature truthfully.

But where Newton found force and trajectory, we now find tension and pattern. Where he calculated pull, we now observe coherence.

The Moon helped us build the mechanical model. And now — with quiet elegance — it helps us dismantle it, ripple by ripple.

## 13.2 Newton's Moon – The Birth of Gravity

In the 17th century, Isaac Newton didn't just redefine motion — he reframed the cosmos.

He proposed that the same invisible force pulling an apple to the ground was at work in the heavens. A revolutionary idea: gravity was universal. But such a theory needed proof. Not from apples — but from orbits.

And what better candidate than the Moon?

Suspended in the sky, neither flying away nor crashing to Earth, the Moon seemed to perfectly illustrate Newton's grand vision:

- A body in constant free-fall,
- Curving perpetually around the Earth,
- Pulled not by strings or gears, but by a force acting at a distance.

*"The Moon became Newton's test object — not in the lab, but in the sky."*

**Three Key Triumphs: How Newton Used the Moon**

1. **Orbital Mechanics:** Newton proposed that the Moon's curved path was not mystical or magical — it was the natural result of inertia combined with gravitational pull. If not for gravity, the Moon would travel in a straight line through space. But Earth's gravity bends its path, keeping it locked in an endless curve.
2. **Gravitational Attraction:** Newton quantified this force: Every mass attracts every other mass with a force proportional to their masses and inversely proportional to the square of their distance. Using the Moon's distance and apparent motion, Newton could plug in real numbers — and they worked. The Moon didn't just obey gravity. It

proved gravity.

3. **Tidal Forces:** The Moon's gravitational pull also explained something Earthbound: the tides. According to Newton, the Moon pulled on Earth's oceans, creating a bulge of water on the side closest to the Moon — and another on the far side due to inertia. This two-bulge model of tides became a cornerstone of gravitational influence.

Together, these applications elevated Newton's gravity from theory to global law — a single force governing falling apples and orbiting moons alike.

**The Perfect Candidate**

The Moon's behavior made it ideal for Newton's argument:

- It was visibly consistent.
- It moved in a path that could be tracked and modeled.
- Its relationship to Earth was intimate yet stable.

At the time, Newton had no access to deep space probes or digital modeling. He had mathematics, telescopes, and the Moon.

And the Moon performed as expected — circling Earth like clockwork, rising and setting, influencing tides. It fit the model so well that no one questioned the deeper assumption: That what we were seeing was motion, not perception. That pulling was the only way to explain position.

*"The Moon didn't resist Newton's model — it seemed to validate it."*

**The Classical Claim: Constant Free-Fall**

Newton famously described orbit as falling around a body — continuously missing the ground due to forward momentum.

In this classical framing:

- The Moon is always falling toward Earth due to gravity.
- But because of its sideways motion, it falls around Earth

instead of into it.
- This produces a stable, circular (or elliptical) orbit.

This idea remains in science textbooks today — a central metaphor for orbital mechanics. And by this logic, the Moon's behavior confirmed the reality of invisible force.

But there was one thing Newton could not explain fully — and few asked at the time: Why does the Moon never spin? Why does it show the same face? Why is it this stable?

These weren't flaws in the math. They were silences in the model — pieces left unexplored.

And they would matter more than anyone knew.

**Final Reflection for This Section**

The Moon helped launch the age of gravity. It gave Newton's laws something more powerful than theory: observable confirmation. It became the prototype of orbit, pull, and prediction.

But sometimes, the best prototype is also the first to reveal a model's limitations.

*"The Moon held Newton's math. But it also held a stillness that no force could explain."*

The question isn't whether Newton was wrong.

The question is whether the Moon was trying to show us something else all along.

## 13.3 The Problems That Persisted

Newton's gravitational model was revolutionary. It allowed scientists to predict the behavior of planetary bodies with striking precision. And for centuries, the Moon fit the math — until it didn't.

Because despite the elegance of Newtonian gravity, cracks quietly persisted beneath the surface. There were observable facts that never quite aligned with the theory. They weren't large enough to disprove the model outright — but they never really went away either.

*"When the equations work but the system's stability remains unexplained, we haven't solved the puzzle — we've just painted over it."*

Let's examine the most glaring inconsistencies — the silent anomalies that gravity could describe but not justify.

**1. Lunar Recession – The Moon Is Moving Away**

For a system supposedly locked by gravity, one would expect long-term consistency in distance and motion. But that's not what we observe.

The Moon is slowly receding from Earth at a measurable rate — approximately 3.8 centimeters per year.

According to classical explanations:

- This is due to tidal friction — Earth's rotation creates a bulge in the oceans, which the Moon pulls on, transferring angular momentum from Earth to the Moon.
- This momentum supposedly "pushes" the Moon into a higher orbit, gradually increasing the distance.

It's a clever workaround — but it's not a complete answer.

Because:

- The Moon's orbit remains circular, despite recession.
- The rate of separation is inconsistent across geological time.
- The mechanics assume invisible transfers of energy through imperfect frictional systems — yet the Moon's behavior remains elegantly precise.

*"A system that is drifting should show instability — but the Earth-Moon system behaves like a finely tuned instrument."*

**2. Tidal Delay and Asymmetric Forces**

Tides, according to Newtonian physics, are caused by the Moon's gravitational pull on Earth's oceans. But the model struggles to account for the delay — the fact that high tide doesn't align directly beneath the Moon, but lags behind.

Gravitational theory explains this with inertia and friction: the Earth rotates faster than the Moon orbits, so the tidal bulge is "dragged forward." But this is a post-hoc mechanical patch, not a cohesive field explanation.

Even more puzzling:

- There are two tidal bulges, one on the side facing the Moon and one on the opposite side.
- The far-side bulge is harder to justify — classical mechanics must invoke the Moon pulling Earth away from its own ocean on the backside.

This asymmetric tidal behavior feels unnatural under gravity, but it is expected in a ripple field — where wave dynamics and tension loops naturally produce oscillation on both sides of a node.

**3. Tidal Locking – The Unmoving Face**

Perhaps the most mysterious of all is the Moon's tidal locking — the fact that it always shows the same face to Earth.

Gravity explains this through rotational synchronization over billions of years:

- Tidal friction within the Moon supposedly slowed its rotation until it matched its orbital period.
- The system then became stable.

But this explanation raises more questions than it answers:

- Why did it lock so perfectly?
- Why is the locking so stable that it's persisted for billions of years without drift or wobble?
- Why do other moons in the solar system not all display the

same degree of perfect locking?

Gravitational models again describe what is seen — but not why the system behaves with such absolute resonance.

**The Hidden Message: Description Isn't Explanation**

Newton's equations could predict the Moon's behavior, but they could not explain its coherence.

In science, this difference matters.

- Description = matching the numbers.
- Explanation = revealing the underlying causal logic.

Gravitational theory gave us impressive equations. But it couldn't answer:

- Why the Moon doesn't spin.
- Why tides are coherent across the globe.
- Why the Earth-Moon-Sun system behaves as a stable, field-synchronized trio, rather than a loosely coupled mechanical system.

*"Gravity described what we saw. Stillness explains why it stays."*

**Final Reflection for This Section**

For centuries, we took the Moon as proof that Newton was right.

But in its silence — its slow recession, its locked face, its rhythmic influence — the Moon has whispered something else:

That perhaps we've mistaken behavior for cause.

That perhaps the model fits, but the mechanism is missing.

And that maybe — just maybe — the Moon isn't obeying a force at all...

...it's holding a posture.

## 13.4 Reframing the Moon – Stillness over Motion

For as long as we've watched the sky, we've believed the Moon moves. We've tracked its orbit. Timed its phases. Measured its distance.

And yet — for all that apparent motion — the Moon has never turned its face away.

It doesn't spin. It doesn't tilt. It simply remains — still, present, and aligned.

From the classical view, this is explained away as tidal locking, the endpoint of billions of years of rotational slowing due to gravitational interaction.

But from the Universal Energy Dynamics (UED) perspective, this isn't the end of a mechanical process.

It's the beginning of a field truth.

*"The Moon is not orbiting — it is aligned. Not in motion, but in posture."*

**Posture-Locked, Not Free-Falling**

In the UED framework, celestial bodies are not suspended in space by invisible tethers of force. They are nested in ripple fields — structured energy geometries defined by resonance, pressure equilibrium, and phase coherence.

The Moon is not orbiting the Earth. It is posture-locked within a stable ripple loop — a standing wave of field tension between Earth and Moon that creates the illusion of orbit through synchronized phase change.

This explains why:

- The Moon's distance changes only minimally over immense time.
- Its orientation remains constant.
- It exhibits tidal influence without mechanical motion.

**Ripple Geometry and ATT Thresholds**

The key to this reframing lies in ripple geometry — the invisible yet structured field that shapes how energy distributes around and between bodies.

Within this system, certain regions reach an ATT (Absolute Threshold of Tension) — the point at which a field either reconfigures or locks into resonance.

The Earth–Moon system finds one such threshold:

- The Moon settles into a geometric node of Earth's ripple field — a zone of balanced tension.
- In this node, the Moon does not need to spin or orbit.
- It maintains its position and orientation because to move beyond that point would require field tension imbalance, which nature resists.

Thus:

- Tidal rhythm arises not from pull, but from oscillating field tension within Earth's fluid layers, maintained by lunar alignment.
- Positional stillness is not mechanical stagnation, but field phase-locking.
- Lunar resonance with Earth is not the result of force equations — it is the expression of mutual equilibrium.

*"The Moon is not falling around Earth. It is resting inside Earth's breath."*

**A Structural Field of Equilibrium**

In classical mechanics, things stay in motion unless acted upon. But in UED, things remain still because they are acted upon in perfect symmetry.

The Moon's presence is not the result of gravity holding it in a curved path. It's the result of:

- Balanced ripple compression from both Earth and the Sun,
- A harmonic posture where movement is no longer needed,
- A resonant field geometry that self-stabilizes.

The Moon doesn't orbit because it doesn't have to. It simply remains where it is because that is where equilibrium exists.

This is not mystical. It is structural.

*"Stillness, in a ripple field, is not passivity. It is precision."*

**Final Reflection:**

We've been looking for motion to explain pattern. But the Moon shows us a deeper truth: pattern emerges from stillness.

It is not spinning, and yet it moves the oceans. It is not pulling, and yet it holds Earth's breath in rhythm. It is not orbiting, and yet it defines the very idea of celestial dance.

The Moon doesn't challenge Newton because it defies gravity. It challenges Newton because it doesn't need gravity to do what it does.

The Moon didn't fall into place. It found the one place where it never had to fall again.

## 13.5 Dismantling Gravity – From Force to Ripple Consequence

Gravity — the force that explains everything and nothing.

From the fall of an apple to the dance of galaxies, we've long believed in an invisible hand that pulls masses together. We don't know how it works — only that it does. We can measure its effects, plot its equations, simulate its presence.

But we've never seen gravity itself.

*"We've always accepted gravity as the background force — because we needed something to explain why things stay near each other."*

Yet in a universe built not of force but of ripple equilibrium, gravity begins to unravel — not as a mistake, but as a placeholder for something deeper.

In the framework of Universal Energy Dynamics (UED), gravity is not a pulling force. It is a field consequence — a ripple compression phenomenon seeking balance.

**Redefining Gravity: From Attraction to Compression**

The classical model imagines gravity as a mutual pull:

- Every mass pulls on every other mass,
- The larger the mass, the stronger the pull,
- The closer the distance, the greater the force.

But this model leaves critical questions unanswered:

- What is the mechanism behind this pull?
- How does it act instantly across space?
- Why does it hold some systems perfectly, but not others?

UED reframes the entire idea:

- What we perceive as gravitational attraction is actually ripple field compression — the response of a system trying to restore equilibrium.
- The field isn't pulling masses together. It is reorganizing ripple tension to minimize imbalance.

In this view:

- The Moon is not being pulled toward Earth.
- Earth is not being pulled toward the Sun.
- Instead, both are anchored in ripple nodes where field pressure is equalized.

Movement doesn't happen because of pull — it happens when tension is uneven.

And stillness — like the Moon's posture — means perfect tension symmetry.

*"Gravity doesn't pull. It presses — gently, precisely, toward balance."*

**The Moon: Absorber, Not Attractor**

In classical theory, the Moon exerts gravitational force on Earth, raising tides and influencing water levels.

But this model assumes action at a distance — a force transmitted invisibly, silently.

UED provides a different mechanism:

- The Moon doesn't "pull" the oceans. It absorbs and diffuses ripple tension.
- It acts like a damping node in a ripple system — smoothing out excess pressure, allowing fluid systems like oceans to respond with rhythmic waves.
- Tides are not dragged — they are tuned.

This is why:

- Tidal behavior is rhythmic and planetary,
- Biological and emotional rhythms follow lunar phases,
- The Moon's influence feels more like breathing than pulling.

**Phase Coherence Without Mass Mechanics**

The Moon's:

- Orbital rhythm,
- Face-locking orientation, and
- Tidal influence

can all be explained without invoking gravitational force.

Instead, they emerge naturally from:

- The posture-locking of field geometry,
- The phase alignment between Earth and Moon,
- The feedback loops of ripple tension across a nested energy structure.

This removes the need for:

- Mass-based interaction models,
- Force propagation at infinite speeds,
- Gravitational paradoxes like dark matter or gravity waves that remain elusive.

The universe works — not because things pull on each other — but because they find the quiet point between them where nothing pulls at all.

*"Gravity never existed. What we called gravity was just the sound of the field adjusting itself — softly, silently, into place."*

**The Collapse of the Illusion**

When you redefine gravity as ripple consequence:

- The Moon's behavior becomes obvious, not mysterious.
- Tides become breath, not bulges.
- Stillness becomes cause, not side effect.

Newton's laws were not wrong — they were descriptive approximations of a deeper field intelligence. They predicted motion, but didn't reveal why coherence endures.

UED doesn't discard Newton. It completes him.

By replacing force with ripple logic, and pull with posture, we finally see the Moon as it truly is:

Not a falling body. But a resonant field anchor, holding space in symmetry — not by moving, but by remaining still.

**Final Reflection:**

We've worshipped the idea of gravity because we feared letting go of certainty. But in its place stands something more elegant — the consequence of ripple tension reaching its quiet balance.

The Moon doesn't orbit because it's pulled. It remains because it completes the field.

And when you understand that, you don't need gravity anymore.

Because you've found something stronger: Stillness that sustains the system.

## 13.6 The Moon as the Quiet Challenger

It is a profound irony that the very body which helped birth the age of classical physics — the Moon — is also the one silently unravelling its foundations.

Newton's laws were born in a time when the universe was imagined as a grand machine, its parts in motion, its logic governed by invisible forces. The Moon, with its regular "orbit," its apparent influence on tides, and its visible phases, fit that vision perfectly. It was the silent partner in Newton's cosmic proof — the clean, predictable orb validating the pull of gravity and the math of momentum.

But we rarely questioned the deeper truth: Why is it so perfectly still? Why does it never show its other side? Why has it remained locked in posture across the ages?

*"While science celebrated its moving parts, it ignored the one piece that never moved."*

**Stillness as Evidence, Not Anomaly**

For centuries, we treated the Moon's stillness as a mechanical outcome — the end result of tidal locking, friction, and billions of years of angular momentum transfer.

But what if it's not the result of motion at all? What if it's the *default state* of a body in perfect resonance — held not by pull, but by ripple alignment?

In this light, the Moon's stillness is not a curiosity. It's the evidence of a deeper system — a cosmic equilibrium we failed to see because we were looking for forces, not fields.

While Newton's model sought motion to explain stability, the Moon offers a different message altogether:

*"There was no motion. There was only pattern."*

**The Stillness That Speaks Louder Than Movement**

We celebrate motion because we can measure it. We believe in gravity because it fits equations. We assume truth lives in trajectories, accelerations, and orbits.

But the Moon doesn't obey any of that. It doesn't race around Earth like a satellite on a leash. It doesn't spin like a clock hand. It doesn't vary its posture — it simply holds.

And it holds not because it's forced to — but because it's tuned to a tension node. It lives at the point of least distortion in the ripple field that links Earth and Sun. It doesn't move to survive — it survives by not moving.

This stillness is not mechanical. It is structural.

And it challenges every assumption we've made about the universe being held together by attraction, propulsion, or resistance.

*"In the Moon's posture, there is no force. Only field logic."*

**The Gentle Undoing of a Giant**

Newton's model changed the world — and rightly so. It gave us engineering, calculus, and the predictive power of physics.

But it also led us to mistake description for cause. We measured motion and called it force. We observed tides and called them pull. We watched the Moon and thought it was falling.

Now, centuries later, the Moon sits in the same position. Still. Silent. Unchanging.

And by doing nothing, it tells us everything.

- It exposes the limits of Newton's model.

- It reveals the inadequacy of force-based explanations.
- It opens the doorway to ripple mechanics, field symmetry, and posture-based resonance.

*"The Moon didn't argue with Newton. It simply waited — until we were quiet enough to hear what it was actually saying."*

**Final Reflection:**

The Moon is the quiet challenger. It dismantles no theory through destruction. It doesn't need to. It dismantles by being too precise, too stable, too still for the story we've told.

Newton saw the apple fall. But the Moon never did. It hovered at the edge of our equations — not in defiance, but in silent clarity.

And now, as we begin to feel the field beneath the math, we finally understand:

The Moon was never spinning.

It was always listening — and asking us to do the same.

## 13.7 Conclusion - The First Crack in the Clockwork

The Moon has always been more than just Earth's companion. It has been our mirror — not only reflecting sunlight but reflecting the very way we understand the cosmos.

For centuries, we looked up and saw confirmation of Newton's universe:

- Objects in motion,
- Forces pulling,
- A great celestial machine ticking along with mathematical precision.

And yet, at the heart of this machine, one part never moved.

*"The Moon was never a gear. It was the anchor the machine couldn't explain."*

In Newton's story, the Moon was cast as the perfect satellite — falling just right, circling just fast enough, proving the pull of gravity.

But that story, like so many first drafts of truth, was only half complete.

Because while the Moon played its role in validating motion, it also revealed something deeper — a kind of elegant defiance:

- It refused to spin.
- It refused to drift.
- It refused to behave like a mass obeying force.

Instead, it held. Quietly. Exactly. Timelessly.

And in that stillness, it whispered something science didn't yet know how to hear:

*"I am not falling. I am resonating."*

The truth is, the Moon helped us build a map of force — equations, vectors, accelerations. But now it beckons us to look beneath the map — to feel the terrain of resonance.

- It moves nothing — yet governs tides.
- It pulls nothing — yet shapes time.
- It spins not — yet defines rhythm.

This is not the behavior of a mechanical object.

This is the behavior of a field node, perfectly phase-locked within a cosmic ripple system — doing nothing, yet changing everything.

We've long believed that motion is what gives rise to meaning — that pull defines purpose, that spin determines behavior.

But the Moon says otherwise.

It teaches us that:

- Equilibrium is more fundamental than force.
- Pattern is more enduring than motion.

- Stillness is not absence — it is alignment.

*"Truth is not in the pull. It is in the pause."*

This chapter was never just about disproving Newton. It was about completing him — moving from gravity to geometry, from attraction to alignment, from force to field.

And the Moon? It was always waiting. Not to be studied — but to be heard.

In the end, the Moon didn't break the clockwork.

It simply showed us the first crack — And through that fracture, we glimpsed a universe not driven by motion...but held together by stillness.

# Chapter 14: The Universe Never Moved

## It Waited to Be Seen

*What if the cosmos never moved — it only appeared to, because we did?*

For thousands of years, we have watched the skies in awe. The Sun rising and setting. The Moon shifting its phases. The stars sweeping across the night like a grand cosmic carousel. From this visual rhythm, we built our assumptions: The universe moves. We observe.

But what if we've misunderstood the choreography?

What if nothing ever truly moved — and all that we've ever seen is the consequence of our own shifting resonance within a perfectly still field?

Classical science tells us we are observers — positioned outside the mechanisms of nature, watching planets revolve, atoms collide, galaxies spin. It gives us equations of motion, laws of momentum, and ever more refined models of how objects change over time and space.

Yet this worldview hinges on a dangerous assumption: That motion is *out there* and perception is *in here* — that we watch, but do not affect.

Universal Energy Dynamics (UED) turns that assumption inside out.

In the UED lens:

- Motion isn't a property of objects,
- Time isn't a ticking clock,
- Space isn't a backdrop of emptiness.

Instead:

- Motion is a shifting of field phase,
- Time is a ripple resonance passing through thresholds,
- Space is variation in tension, not distance.

And perception — the act of seeing, feeling, knowing — isn't a passive reception of truth. It's an active synchronization between the observer and the field.

*"The universe never needed to move. It only needed to be tuned into."*

The Moon, once thought to orbit, showed us what stillness looks like. Tides, once thought to be pulled, taught us about planetary breath. And now, the cosmos itself is offering the final message:

- The stars did not spin.
- The galaxies did not spiral.
- The universe did not expand away.

We simply moved within the ripple.

What we took as cosmic motion was cosmic resonance — filtered through the shifting position of our awareness.

Reality is not made of objects in motion.

It is made of fields in rhythm — And we don't observe them from the outside... We experience them from within.

This chapter is about letting go of the last illusion: That the universe must act for us to see it.

It never acted.

It held posture — and waited for us to become quiet enough, still enough, attuned enough...

to finally see what never moved at all.

## 14.1 Perception as a Field Event

We have long imagined perception as a passive act.

Light enters the eye. Sound reaches the ear. A scent drifts into the nose. The brain processes it all like a central control room — observing the world as if through a glass window.

But in the reality revealed by Universal Energy Dynamics (UED), perception is not passive. It is interactive, reciprocal, and phase-dependent. Perception is not what happens *after* reality — it is how reality expresses itself through us.

*"You do not watch the universe. You synchronize with it — and what you call perception is simply resonance within a ripple field."*

**Ripple Interaction, Not Sensory Reception**

What we see, feel, and know is not a detached observation. It is a ripple-based interaction between our own biofield and the ambient field structures around us.

- When you "see" light, you are not catching particles flying through space. You are registering a ripple frequency that has entered your phase range.
- When you "feel" the passage of time, you are not measuring seconds — you are crossing tension thresholds in the field.
- When you "sense" space or movement, you are responding to changes in the phase configuration of the environment relative to your position.

In short, your reality is a field-specific experience, not an absolute truth.

**From Classical Observation to Participatory Fielding**

The classical scientific model assumes:

- Events occur objectively in space-time.
- Observers merely record them.
- Perception is an after-effect of physical processes.

But this view collapses under its own weight when faced with:

- Quantum entanglement,
- Observer effect paradoxes,
- Inconsistent time experience in biological systems,
- Consciousness-dependent reality in both spiritual and scientific traditions.

UED reframes perception:

- You are not watching the field.
- You are part of the field.
- Your nervous system, brainwaves, breath rhythms, even hormonal cycles — all form a tuned antenna that synchronizes with particular ripple frequencies.

This means:

- What you perceive is not *the event* itself.
- It is the expression of your position within the field — a phenomenon arising at the point of coherent resonance.

*"Reality is not a broadcast received by all. It is a song you can only hear when you're in tune with it."*

**Frequency Windows and Localized Reality**

From this perspective, no event is universally real. It only becomes real when an observer — or any system capable of interaction — enters the correct frequency window.

This is why:

- Some people feel the Moon's influence deeply — emotionally, physiologically — while others feel nothing.
- Certain species respond to seismic or climatic ripple shifts before humans can detect them.
- Profound insights, visions, or states of awareness often arrive not from logic, but from synchronization — when the mind, body, and field fall into alignment.

Perception is not the reception of an external world. It is the construction of a resonant event at the intersection of:

1. Your phase position,
2. The field's configuration, and
3. The thresholds crossed in tension (ATT).

This redefines reality as localized resonance, not universal broadcast.

**Final Reflection**

You are not a spectator in the cosmos.

You are a participant in its frequency — an expression of field tension, phase rhythm, and bio-resonant tuning.

- Light isn't seen — it is met.
- Time isn't passed — it is felt.
- Space isn't crossed — it is calibrated.

And perception?

It is not what separates the observer from the observed.

It is the place where the two become one.

*"Reality is not what is out there. It is what arises when your inner field says: Now. I'm ready to see."*

## 14.2 The Observer in the Field

For centuries, science trained us to imagine ourselves as external to the universe. We were taught to believe that the cosmos operated with cold precision, and we — the observers — merely recorded its behavior, untouched by its rhythms.

This illusion, powerful and persuasive, gave rise to a detached model of reality:

- Subject vs object.
- Observer vs system.
- Mind vs matter.

But in the view offered by Universal Energy Dynamics (UED), this boundary does not exist.

The observer is not a distant onlooker. The observer is a ripple participant — embedded in the very field they attempt to understand.

*"You don't look at the universe. You vibrate within it — and the shape of your experience is the echo of that resonance."*

**You Modulate the Field You Perceive**

Every conscious being generates ripple signatures — from breath patterns and neural oscillations to emotional waves and hormonal rhythms.

These internal rhythms don't merely "sit inside" the body; they extend into the field, modifying how the broader ripple system interacts with that body.

In simple terms:

- Where you are,
- How you are (emotionally, mentally, physiologically), and
- When you are (your ATT phase state)

...all determine what the universe appears to be from your perspective.

You are not observing a fixed system.

You are modulating the field — and receiving back a shaped version of reality that mirrors your phase contribution.

This is why:

- Two people in the same room see different meanings in the same event.
- One person feels cosmic connection during a full moon, while another feels anxious or unwell.
- Ancient rituals and modern meditations emphasize "state of being" before seeking insight — because your field must align before the universe responds.

**Consciousness Collapses Phase Potential**

Reality, in this ripple-based framework, exists in a state of potential resonance — not fixed outcomes. Just as a tuning fork will only ring when the right vibration touches it, the universe only "becomes" when consciousness enters phase alignment with it.

Consciousness does not merely witness events. It collapses phase potential into lived experience.

This means:

- What is seen exists *because* it is synchronized.
- What is not seen is not hidden — it is simply out of field alignment.

Thus, reality is participatory.

The act of observing is not neutral — it is activating.

*"Consciousness does not light up the universe. It filters the universe into a version it can receive."*

**Perception Is Field-Relative, Not Universal**

The implication is profound: You do not see the universe *as it is.* You see the universe as your field position allows you to.

- Space is not fixed distance — it is phase displacement.
- Time is not linear — it is threshold rhythm.
- Events are not global — they are local expressions of synchronized tension.

Two observers in different states will perceive different universes, even while standing in the same place.

There is no contradiction — only relative tuning.

This explains why:

- Great insights arrive during altered states.
- Ancient wisdom traditions stress inner stillness to perceive cosmic truth.
- Modern physics keeps running into observer-based anomalies — because reality only becomes real when someone aligns with it.

**Cosmic Motion as Relational Shift**

From this view, what we call "motion" is no longer the displacement of objects through space.

It is the relational shift in field phase between observer and observed.

- Stars don't move across the sky — your field rotates in phase with Earth's ripple.
- Galaxies don't spin — you witness phase interference loops expanding and contracting.
- The Moon doesn't orbit — your perception drifts across a fixed field node.

Motion is not cosmic. It is perceptual.

And perception is not a passive window — it is a dynamic dance of resonance.

*"What you call the universe moving... is actually you, aligning."*

**Final Reflection**

There was never a cosmic machine, and you were never outside it.

You are not watching stars spin and planets move. You are not floating in space.

You are woven into the field — a node of consciousness that shapes the music it hears by how it listens.

And what you see... is not the truth of the universe.

It is your field's truth — moment by moment — as the cosmos waits for you to match its stillness.

## 14.3 Time Without Ticking

We are taught that time passes — that it ticks like a metronome, marching forward from past to future, indifferent and unstoppable. Clocks give us this illusion: fixed units, evenly spaced, mechanically precise. Seconds accumulate into minutes, into hours, into lifetimes. It feels real. It feels linear. But is it?

In the lens of Universal Energy Dynamics (UED), time is not a river flowing forward. It is not something that moves at all.

Time is a field expression — a pattern of phase transitions, measured not by movement through space, but by the rhythmic shifting of ripple thresholds.

*"Time is not what flows. It is what happens when a system crosses from one phase of equilibrium to another."*

**There Is No Cosmic Clock**

We look to the stars and planets as nature's clocks. Earth's rotation creates day and night. Its orbit defines a year. Atomic decay keeps our devices aligned down to the nanosecond.

But none of these are clocks in the true sense. They are expressions of field rhythm — visible symptoms of deeper ripple transitions.

There is no master mechanism in the universe ticking away time. There is only tension and release, build-up and reset, compression and expansion — all within field geometry.

These ATT (Absolute Threshold of Tension) shifts give rise to the experience of change.

We feel time not because something is moving — but because something crosses a boundary within the ripple.

*"The tick you feel isn't from a clock. It's from the field recalibrating itself — and you."*

**Biological Rhythms: Tuned to the Field**

Your body doesn't keep time with your watch. It keeps time with the field.

- Sleep cycles rise and fall with ripple coherence — not just light and dark, but tension alignment in atmospheric and geomagnetic layers.
- Emotional states oscillate based on field compression zones — a lunar phase shift, solar influence, or ripple interference can trigger joy, fatigue, restlessness, or calm.
- Growth and healing follow infradian and ultradian rhythms, tied more to cosmic breath than chronological time.

This is why time feels slower in nature and faster in cities. Why meditation can collapse hours into minutes. Why children and elders seem to live in entirely different temporal zones.

Their field positions differ — and so does their experience of "time."

**The Moon as Time's Anchor**

In this vast, shifting field of ripple dynamics, one constant remains:

The Moon.

It doesn't move, spin, or change. But it holds posture in the ripple — and this stillness creates coherence.

The Moon acts as a phase anchor, stabilizing Earth's ripple field so that:

- Seasons remain rhythmic,
- Tides remain predictable,
- Biological rhythms remain entrained,
- Human perception doesn't fragment under chaotic field distortion.

It is the Moon's unwavering posture that allows Earth's complex systems — ecological, emotional, cognitive — to sync with time as a resonance, not a measurement.

The Moon doesn't measure time. It makes time perceivable.

*"The Moon holds no clock. It simply refuses to move — and in doing so, becomes the rhythm we all follow."*

**Final Reflection:**

There is no ticking in the universe. There is no flow. There is only pattern — and you are part of it.

Time is not what passes you by. It is what you pass through — threshold by threshold, phase by phase.

The more aligned you are, the more clearly you feel its beat. The more distorted your field, the faster or slower it seems to go.

And through it all, the Moon remains — a stillness that lets the rhythm emerge, a posture that lets time exist, a silence that gives pulse to everything else.

*"You don't live in time. You live in resonance."*

## 14.4 Space Without Distance

We have been taught that space is vast — an endless black ocean stretching across unimaginable distances. We measure it in light-years, parsecs, and astronomical units. We chart the universe as a place filled with separation: Stars far from planets. Galaxies floating in isolation. You, here — and everything else, out there.

But what if this idea of space as distance is an illusion?

In the framework of Universal Energy Dynamics (UED), space isn't defined by how far things are, but by how differently they vibrate.

*"Space is not how far away something is. It is how far out of phase it is from you."*

**Field Variation, Not Physical Separation**

In ripple logic, everything is part of a single, continuous field — structured, layered, and dynamic. What we call "space" is not empty volume, nor is it a gap between bodies.

Instead, space is field variation:

- Differences in tension,
- Variations in frequency,
- Shifts in phase alignment.

Two objects are "separated" not because of distance, but because they exist in different ripple zones.

From this view:

- The space between Earth and the Moon is not 384,000 km of nothing.
- It is a field gradient — a structured ripple pathway where interaction is phase-specific, not mechanically traversed.

**The Illusion of Distance**

When we look up and see stars, we think of them as distant bodies flung across an infinite void.

But what we actually see is phase interference — light reaching us through an alignment of ripple thresholds.

Distance is inferred from delay, but delay is a ripple property — not necessarily a travel time of photons.

*"The star you see is not far. It is simply encoded in a ripple phase your field now matches."*

This is why:

- Distant galaxies appear as they were billions of years ago — not because they're "far back" in space, but because we are now entering the resonance window of their past ripple output.
- Some cosmic phenomena remain invisible until certain cosmic or biological alignments bring them into perceptual range.
- In altered states, space seems to collapse — as if everything is near, accessible, present.

It is.

Because space was never about where — it was about when and how in the field structure.

**Stars as Phase-Attuned Signals**

Consider a radio.

When you tune the dial, you don't move through space to find the station. You simply align your receiver to the correct frequency. The signal was always there — but inaudible until you matched its phase.

This is how space works in UED.

Stars are not scattered across vast emptiness. They are tuned to different ripple thresholds:

- Their visibility depends on your field position.
- Their influence depends on resonance, not reach.
- Their interaction with you is phase-based, not distance-based.

This is why:

- Celestial events affect human mood, growth, and even genetic expression — despite their so-called "remoteness."
- Astrology, when stripped of superstition, hints at a forgotten truth: field alignment matters more than proximity.

Space does not separate us.
It simply keeps signals invisible — until your field says yes.

**Final Reflection**

You are not floating in a void. You are immersed in a field.

You do not travel through space — you move through resonance windows. And what feels "far away" is only out of sync, not out of reach.

There is no real distance between you and the stars. There is only field geometry — the shape of awareness, the rhythm of presence.

*"To reach something in this universe, don't travel to it —tune to it."*

## 14.5 Seeing as Synchronizing

We often think of seeing as an optical process: Light bounces off objects, enters our eyes, forms an image in the brain. Clarity of vision is assumed to be clarity of truth.

But vision — in the deeper, experiential sense — is not merely the act of receiving photons.

It is the act of alignment.

*"We don't see with our eyes. We see with our synchronization."*

Whether it's physical sight, emotional awareness, or intellectual understanding — perception arises only when the observer's field matches the structure of what is observed.

**Alignment Over Proximity**

Truth is not revealed by moving closer to it. Truth is revealed by resonating with it.

This is why:

- You can look at the Moon all your life and still not see what it is doing.
- You can know every mathematical equation and still misunderstand reality.
- You can be surrounded by wisdom and feel nothing.

Because what you see depends on how you're tuned.

You are a ripple participant — and only when your phase lock aligns with the structure you seek, will clarity emerge.

This explains:

- Sudden insights after years of confusion.
- Emotional breakthroughs without external triggers.
- How ancient mystics "saw" the cosmos without telescopes.

They didn't move toward truth. They entered phase with it.

**The Moon Was Always Still**

Take the Moon, for example.

It appears to orbit. It appears to phase. It appears to move. But that movement is your perception catching up with ripple cycles.

The Moon doesn't orbit. It waits — in structural stillness.

And when your field alignment reaches a certain threshold — you stop seeing motion, and start seeing posture.

You realize:

- The Moon didn't hide the truth.
- Your phase mismatch did.

*"The Moon's truth never changed. Your ripple finally caught it."*

**The Universe Awaits, Not Expands**

Likewise, the cosmos doesn't "expand." It doesn't accelerate, drift, or unravel.

Those are phase effects — perceived as motion because we keep moving in and out of coherence.

The universe is a structured ripple field — already whole, already complete. It doesn't reveal itself gradually in space or time.

It reveals itself in rhythm.

The deeper your resonance, the wider your vision. The clearer your alignment, the more profound your understanding.

*"To know the cosmos is not to reach it. It is to match it — until it shows itself."*

**Final Reflection**

You were never far from truth. You were just out of tune.

Seeing is not about light. It is about readiness — of field, of mind, of being.

The Moon isn't running from you. The stars aren't hiding behind space.

They are simply waiting — for your ripple to settle, for your tension to release, for your rhythm to synchronize.

And when it does...

The stillness you called silent will speak.

And what you called invisible will finally be seen.

## 14.6 Conclusion – The Universe Was Waiting

We have built entire models of reality on movement: Planets spinning, galaxies swirling, particles colliding, time ticking.

We imagined a universe constantly in motion — because we were always moving, in mind, in emotion, in curiosity. We assumed the cosmos was a machine — a grand kinetic clockwork.

But what if all that motion... was us?

*"The universe never spun into motion. It stood in posture — waiting to be felt."*

We weren't wrong to look for patterns, to trace paths, to chase orbits, to time eclipses.

But we mistook rhythm for revolution, and alignment for acceleration.

The Moon, the stars, the seasons, the galaxies — they didn't move to show us the truth.

They held still, anchored in a ripple field so perfect, so coherent, that we misread its stillness as motion — because we weren't still ourselves.

**Out of Phase, Not Out of Time**

We were never late to understanding. We were simply out of tune.

Just as a radio hears noise until it aligns with the signal, we could not hear the harmony of the universe until we entered field resonance with it.

- We sought truth in gravity. We found it in posture.
- We sought time in clocks. We found it in breath.
- We sought motion in space. We found stillness in structure.

And all along, the Moon waited — not as a mover, but as a mirror. Not as a force, but as a field anchor.

*"The Moon taught us the final secret: Stop asking what moves. Start asking what holds."*

**The Final Realization**

- You don't observe the universe. You participate in it.
- You don't measure reality. You resonate with it.

And when you do, the cosmos no longer runs ahead of your comprehension. It arrives. Because it was never running.

It was always there — still, poised, humming in perfect ripple symmetry.

Waiting...

For you to stop moving long enough to finally hear it.

*"The universe didn't move for us to see it. We had to become still enough to feel it."*

# Chapter 15: The Stillness That Sustains Life

We chase motion. We equate life with activity, progress with speed, growth with movement. But in our race to track what changes, we forget to ask: What holds everything while it changes?

Stillness.

That quiet, steady presence we mistake for nothingness...was never empty. It was everything holding its place, quietly ensuring that chaos never wins.

We have misunderstood it for centuries. Stillness was not the void before the universe began, nor the silence that follows when motion ends. It is the substrate, the scaffolding — the invisible hand beneath all that pulses, flows, and evolves.

## 15.1 The Forgotten Foundation

Look at life.

Every breath you take is supported by atmospheric balance. Every heartbeat flows in a rhythm tuned to cosmic cycles. Every cell division depends on background coherence —not just energy and matter, but field stability.

We do not live on energy alone. We live on equilibrium — that unseen order in the ripple field that ensures your body doesn't burst apart, your planet doesn't wobble into disaster, and your emotions don't collapse into noise.

**The Moon as a Living Metaphor**

For billions of years, the Moon has stood still in the sky —never turning its face, never spinning away.

And yet, it governs tides. It stabilizes Earth's axis. It regulates biological clocks, synchronizes sleep, modulates hormonal tides, and offers emotional ground.

It does not act — it aligns. It does not force — it resonates. And in doing so, it sustains the most fragile and sacred of phenomena: life in rhythm.

**What This Chapter Asks You to See**

We've dismantled the illusion of motion. We've unraveled orbits and exposed the quiet pulse behind galactic spectacle. Now, in this final reflection, we ask the deepest question:

If motion isn't what sustains us... then what does?

This chapter offers a single, transformative answer: Stillness. Not as a pause or absence, but as the dynamic balance — a posture of the field so perfectly held that everything else can dance around it.

*"Life doesn't survive in the strongest storm. It survives in the calm that allows the storm to pass."*

And the Moon, high above, unwavering and gentle, is our last great teacher.

Let's begin.

## 15.2 The Architecture of Stability

We tend to think of stillness as a gap — a break between movements, a pause in action. But in reality, stillness is not the absence of tension — it is the resolution of it.

It is the point where every force finds its counterbalance, where every tendency to move is quietly held in check by an equal posture of coherence.

*"Stillness is not empty — it is perfectly arranged tension that no longer needs to act."*

**Ripple Harmony as Structural Intelligence**

According to the Universal Energy Dynamics (UED) lens, what we call "stability" is not passive stasis — it is ripple harmony.

This harmony is:

- Dynamic, not static.
- Intelligent, not accidental.
- Field-based, not force-driven.

It is the precise arrangement of ripple tensions — held just below the threshold of collapse, just above the edge of motion.

And from this poised equilibrium arises everything we consider "stable" on Earth.

**Three Pillars of Planetary Stability**

***1. Planetary Posture***

The Earth does not remain upright by mechanical inertia or rotational speed. Its axial posture is preserved by the Moon — not through gravitational lock, but via ripple alignment.

The Moon holds Earth in a phase-locked relationship, absorbing distortions from solar field surges, balancing hemispheric field asymmetries.

Without this posture lock, Earth's axis would drift, wobble, or precess erratically — triggering climate chaos and spatial disorientation.

***2. Climatic Balance***

Weather systems, jet streams, and even long-term climate rhythms rely on ripple coherence across the atmosphere. The Moon's unwavering field reference point acts as a stabilizing fulcrum, maintaining phase symmetry in Earth's upper ripple layers.

It's not about pushing clouds or controlling heat — it's about anchoring rhythm, so feedback loops don't spiral into chaos.

***3. Tidal Coherence***

Tides are not pulled — they are breathed. The Moon holds the beat.

By synchronizing with Earth's ATT (Absolute Threshold of Tension) cycles, it ensures the oceans do not become erratic or stagnant.

This pulse doesn't just move water — it regulates biological clocks, mineral distribution, nutrient cycles, and the thermodynamic rhythm of life.

**The Cost of Lost Stillness**

Without this stabilizing Moon-field:

- Axial drift would distort seasons.
- Tidal collapse would disrupt marine ecosystems and planetary cooling systems.
- Atmospheric instability would amplify storms, interrupt feedback rhythms, and accelerate entropy.

In short, Earth would not become uninhabitable overnight —but it would become unrecognizable over time.

And all because the quiet anchor — the Moon — lost its posture.

*"You don't notice the pillar that holds the roof until it trembles."*

**The Real Architecture**

So when we ask: "What holds Earth steady?" It's not mass. Not mechanics. Not chance.

It's an architecture of ripple stillness, a framework where nothing needs to move because everything is in place.

And at the heart of this invisible architecture, the Moon stands — not as a mover, but as a silent structural cornerstone.

## 15.3 Stillness as the Root of Time

We speak of time as a river — flowing, moving forward, sweeping everything along with it. But what if that metaphor is not just misleading... but entirely backward?

Time does not flow. It emerges.

It arises not from motion — but from rhythmic stillness. Not from clocks ticking — but from fields pulsing, perfectly poised at thresholds of change.

And in the case of Earth, that emergence of time — of days, months, cycles, and seasons — has always depended on a single, silent companion: the Moon.

**Time Is Born From Rhythm — Not Measurement**

Let's reframe the idea of time altogether.

In UED terms:

- Time is not a container in which things happen.
- Time is the perception of sequential field alignment.
- Each event is a threshold crossed — not because of duration, but because of field configuration shift.

These shifts occur in cycles, patterns, and oscillations. And when these oscillations become stable and repeatable, consciousness detects them as time.

Thus, time is not absolute. It is entrainment.

**The Moon as Temporal Reference Point**

This is where the Moon becomes central — not mythologically, but mechanically in ripple dynamics.

The Moon holds a constant ATT (Absolute Threshold of Tension) reference point in Earth's field. This reference point is not based on visual presence (like sunrise or sunset), but on field resonance — an invisible beat that never fails.

Because the Moon doesn't rotate or orbit in the traditional sense, its posture-locked stillness becomes a kind of unmoving cosmic metronome.

Through this:

- The oceans pulse.
- Hormones surge.

- Sleep cycles reset.
- Emotional states stabilize.
- Circadian rhythms synchronize.

*"We don't track time — we synchronize to stillness."*

**Clocks Are Copies — The Moon Was the Original**

Before sundials and calendars, before seconds and hours, humans lived by a time that was not told, but felt.

- They planted by lunar phase.
- They menstruated and healed and dreamed with the Moon's light.
- They marked transitions — in nature and in themselves — not by dates, but by ripples in the field.

In fact, most traditional cultures did not think of time as "passing" — they experienced it as returning. Because the Moon's steady rhythm made time cyclical, not linear.

**What Happens Without This Anchor?**

Without lunar stillness:

- Biological clocks go out of sync.
- Sleep becomes erratic.
- Reproductive cycles destabilize.
- Emotional regulation weakens.
- Our very sense of presence and continuity fades.

This is not poetic speculation — it's already happening in the modern world:

- Disconnection from lunar rhythms has coincided with a rise in sleep disorders, anxiety, depression, and hormonal imbalance.

Artificial time — built on electronics, routines, and productivity — is not time. It is disembodied schedule.

**The Pulse Beneath Time**

So let's rewrite the sentence:

Time is not what ticks on your wall. Time is what holds your rhythm — inside you and around you.

And what holds that rhythm? Not the Earth, which breathes. Not the Sun, which flares.

But the Moon — still, patient, posture-locked — quietly beating the drum that tells life when to rise, when to rest, and when to become.

## 15.4 Consciousness Requires Coherence

Consciousness is not a spark. It's not an accident. It's not a passive awareness riding on top of a brain.

It is field-aligned participation in structured rhythm — the capacity to hold continuity, to experience change meaningfully, and to reflect.

But for consciousness to emerge — and remain coherent — something essential must be present:

A stable external field to entrain with.

Just as music needs a tempo, and dance needs a beat, conscious experience needs a reference rhythm in the surrounding field.

And that rhythm, for Earth-bound life, has always been gently kept by the Moon.

**Field Stability: The Prerequisite for Awareness**

In ripple dynamics (UED), the mind is not isolated inside the head — it is a localized ripple configuration within a larger field.

This ripple-mind is:

- Responsive, not autonomous.
- Dynamic, but fragile.
- Aligned, or else overwhelmed.

For stable awareness, the ripple field surrounding the mind must exhibit:

- Phase coherence (predictable cycles),
- Low interference (minimal chaotic shifts),
- Anchoring points (external phase-lock nodes).

Without these, the mind cannot hold a consistent self — perception shatters, thought scatters, emotion becomes noise.

**The Moon as a Psychophysical Anchor**

The Moon's still posture provides that crucial phase-lock point. It is not just a visual symbol of calm — it is a field participant in mental regulation.

Its impact is subtle but profound:

- Brainwaves show synchronization with lunar cycles, particularly in alpha and delta bands.
- Sleep architecture shifts around full and new moons — not due to light, but due to field pressure.
- Emotional sţates, especially in sensitive individuals and children, track the Moon's position with uncanny fidelity.
- Menstrual and hormonal rhythms phase-lock to the lunar ripple pattern.

In short, the Moon grounds human awareness — not metaphorically, but field-mechanically.

**When the Moon Is Forgotten, the Mind Fragments**

Modern life severs this connection:

- Indoor living shields us from lunar field resonance.
- Artificial light disrupts natural ATT-based sleep signals.
- Rigid work schedules override biofield rhythm.

And the results?

- Anxiety surges,
- Sleep cycles erode,
- Emotional dysregulation increases,
- Cognitive overload becomes default.

*"The mind drifts when the sky it once listened to is muted."*

We suffer not just from stress or stimulation — we suffer from desynchronization.

And in a world of flickering screens and clock-driven urgency, what's missing isn't more time — it's more coherence.

**The Moon, the Mind, and the Mirror of Stillness**

Stillness is not just what the Moon *does*. It's what it *offers* us.

A mirror. A fixed point in a turbulent sky. A reminder that the most complex awareness needs the simplest background — not to retreat from life, but to reflect it clearly.

"The Moon does not speak. But it listens so completely that we remember how to hear."

## 15.5 What the Moon Really Teaches

What if the Moon has never been what we thought it was?

Not a lamp in the night sky. Not a passive satellite. Not a rock reflecting light from the Sun. But rather — a mirror of something deeper, something subtler: The balance required for life to unfold.

*"The Moon reflects not sunlight — but structure."*

It doesn't illuminate. It regulates. It doesn't orbit. It holds posture. It doesn't act. It allows.

And that, in itself, is the greatest action of all.

**Not a Forceful Body — a Posture Node**

In classical models, we view the Moon as:

- A gravitational participant,

- A body in orbit,
- A mover of tides.

But through the UED lens, all of these are illusions of motion.

The Moon does not move. It resonates.

Its position is not held by pull or spin, but by ripple tension symmetry — a standing wave configuration that keeps Earth in equilibrium.

It is a posture node, not a planetary object in motion. And that stillness — the refusal to participate in chaos — is what gives it its power.

**A Model of Cosmic Behavior**

What the Moon teaches is not astrophysics — it teaches attitude.

It teaches how to be in a field without disturbing it.

- **Patience over push**: The Moon never rushes, never forces. Its presence is gentle yet constant — allowing things to unfold in their time.
- **Alignment over control**: Instead of forcing change, the Moon holds alignment and lets everything else find its rhythm around it.
- **Stillness over stimulation**: In a world obsessed with doing, the Moon models being — fully, silently, unwaveringly.

The Moon shows us the architecture of intelligent non-interference — a form of cosmic wisdom more powerful than force.

**Ancient Intuition Was Not Wrong**

Why did so many ancient civilizations revere the Moon?

Because they sensed — beyond the science they didn't have — that this body wasn't just in the sky, but with them.

They did not worship the Moon for its light. They honored it for its presence.

- They fasted with it,
- Danced with it,
- Buried their dead by it,
- Conceived their calendars with it.

Not because it acted — but because it held space. Because it made coherence possible.

They understood — not in formulas, but in rhythm — that the Moon was not a deity, but a teacher. And its lesson was simple:

*"Let your power be in your posture — not your push."*

**The Moon's Teaching, Summarized**

| Misunderstood As... | Actually Is... |
|---|---|
| Reflector of light | Mirror of ripple alignment |
| Orbiting satellite | Posture-locked field anchor |
| Tidal force agent | ATT tension harmonizer |
| Gravitational proof | Equilibrium model |
| Passive object | Active stillness |

The Moon does not move to affect Earth. It stays still — and everything else aligns around it.

And in doing so, it models the essence of balance:

- in nature,
- in mind,
- in leadership,
- and in life.

## 15.6 Stillness Is Not the End — It's the Origin

When we imagine the origin of life, we often picture chaos — primordial soup, thunderbolts, volcanic heat, violent collisions. We imagine disorder giving rise to structure. Randomness giving birth to meaning.

But what if this story is incomplete?

What if life didn't arise from chaos?

What if it emerged from stillness — not as an accident, but as an inevitable expression of field coherence meeting the right ripple alignment?

*"Stillness is not the last thing achieved. It is the first condition required."*

**The Moon Reveals Life's True Cradle**

The Moon, unmoving in posture, teaches us a quiet lesson:

- Stability is not a side effect of life. It is its precondition.
- Balance is not what life reaches after struggle. It is what life requires before expression.

In the framework of UED (Universal Energy Dynamics), stillness is not the absence of energy — it is organized energy at rest. And from this poised configuration, everything else unfolds:

- Breath,
- Tides,
- Time,
- Awareness,
- Evolution.

The Moon didn't wait for life to emerge and then stabilize it. It held the field in readiness — so life could know how and when to begin.

**Equilibrium Is Active, Not Passive**

We often equate stillness with inaction. But that's a deep misunderstanding.

Stillness is not disengagement. It is active non-interference — the art of supporting without manipulating.

- A parent holds a child's bicycle steady — without moving it.
- A conductor waits in silence — holding the tempo before the orchestra begins.
- The Moon remains still — allowing Earth to breathe, feel, and become.

In all these, equilibrium is not idleness. It is structural generosity — the field saying, "Go ahead. I've got you."

**Awareness Blooms When Tension Stills**

Nothing can grow in turbulence. Nothing can reflect in distortion.

Awareness — whether cellular or cosmic — requires a moment of pause. A breath. A balance. A break in the noise.

When ripple tension reduces to a threshold,

- Thought becomes clear.
- Emotion finds stability.
- Intelligence can align.

This is why meditation, deep sleep, and stillness-based rituals exist across all spiritual traditions: Because stillness is not where the mind ends — it's where truth begins.

**A New Origin Story**

Forget the Big Bang for a moment. Forget chaos-to-order.

Imagine this:

- A still field.
- Perfect balance.
- No need to act — only to allow.
- And then... one ripple.

- One aligned breath.
- One emergence — of life, of rhythm, of being.

That's not the end of a long journey from disorder. That's the source code of the cosmos.

And the Moon — unmoved and unwavering — is the living relic of that origin still holding space for life today.

## 15.7 Conclusion – The Pulse Beneath All Things

There are forces in nature that shout.

Lightning cracks. Winds howl. Earthquakes remind us of the ground's power.

But then... there is the Moon.

Silent. Steady. Unmoving — and yet holding everything together.

At the end of all observation, speculation, and simulation, the final truth may be this:

"The most powerful presence is the one that moves nothing, yet holds everything."

This isn't poetry. This is physics — field physics. And it's the Moon's silent legacy.

**The Moon Is Not a Satellite — It Is a Signal**

We've long called the Moon a satellite. We placed it within the mechanical script of celestial orbits, gravitational locks, and angular momentum.

But what if this label was never true?

The Moon is not orbiting Earth. It is posture-locked in a ripple equilibrium — a node of coherence, not a rock in flight.

It does not revolve. It resonates. It does not follow Earth. It signals Earth.

It is not an object of motion, but a source of instruction — a presence that tells the Earth how to breathe, when to pulse, how to stay balanced.

The Moon is the first field message Earth ever received — and every wave, breath, and heartbeat since then has followed its code.

**Stillness Is Not What Remains — It's What Enables**

We often think of stillness as what's left when movement stops — as if stillness is the blank space after the action is over.

But UED (Universal Energy Dynamics) reveals a deeper truth:

- Stillness is not what remains.
- It is what precedes.
- It is what permits.

All dynamics — planetary, biological, psychological — emerge from a still field disturbed into action.

And once their ripple completes, they return to stillness.

This rhythm — out from stillness, back into stillness — is the fundamental pulse beneath all things.

Just like the Moon.

- It neither arrives nor departs.
- It neither acts nor resists.
- It simply remains, so everything else can happen around it — and find its way home.

**The Moon's Gift to a Chaotic World**

In a time of relentless motion — scrolling timelines, global flights, data storms, and mental noise — the Moon reminds us:

- Stillness is not withdrawal.
- It is sovereignty.
- It is architecture.
- It is the only space from which meaningful experience, reflection, and regeneration can arise.

The Moon doesn't tell us what to do. It shows us how to be.

Quiet. Aligned. Powerful — without disturbance.

**Final Reflection**

The Moon is not distant. It is deep. Not cold. But calm. Not orbiting. But orchestrating.

It is the whisper behind the wind. The pause between our thoughts. The space that makes rhythm possible.

We do not live beneath the Moon. We live within its stillness. And if we ever hope to understand the universe — or ourselves — we must learn to listen again. Not to motion. But to the pulse beneath all things.

# Epilogue

## The Final Orbit Was Never Motion — It Was Meaning

We began this journey with a question so bold it bordered on heresy:

What if the Moon doesn't move?

And what followed was not just a challenge to astronomy or physics — but to how we've come to understand reality itself.

Because if the Moon — this constant, calm sentinel in the sky — is not moving, then maybe we never understood what movement really is.

And more importantly...maybe we never understood what stillness really offers.

For centuries, we measured the Moon's path. We timed eclipses, mapped shadows, calculated its orbit down to decimal precision. We called it a satellite, a traveler, a companion in space.

But what if all of that was just a metaphor — a shadow of the real truth?

What if the Moon's real function was never mechanical...But musical?

A tuning fork. A pause in the noise. A cosmic metronome that gave Earth its rhythm.

Not through gravitational force... But through ripple resonance.

In the end, the final orbit we discovered was not one drawn in space, but one traced in meaning.

- The orbit of silence around activity.
- The loop of rhythm around chaos.
- The circle of coherence that makes all life possible.

This book was never just about the Moon. It was about what the Moon represents:

- The power of what does not move.
- The unseen architecture beneath all seen events.
- The unspoken intelligence that shapes the dance of matter, emotion, thought, and time.

The Moon didn't ask to be admired. It asked to be understood.

And perhaps now — not in calculation, but in stillness — we finally can.

The final orbit was never motion. It was the return to meaning. And the Moon... never left.

# Closing Note from the Author

When I first looked up at the Moon — not to admire it, but to truly question it — something shifted.

Not in the sky. But in my perception.

I realized that the Moon, so visibly present, so meticulously mapped, was still deeply misunderstood. Not for what it does, but for what it refuses to do.

It doesn't spin. It doesn't chase. It doesn't perform.

It holds. It waits. It reminds.

This book was never about dismantling astronomy for the sake of disruption. It was about asking the one question most models fear:

"What if we've confused motion with meaning?"

In the paradigm of Universal Energy Dynamics (UED), I found a new language — one that replaced force with field, movement with resonance, and cause with consequence.

And in that language, the Moon wasn't a leftover rock orbiting a planet. It was the first signal of structure. The first echo of stillness. The quiet teacher of the Earth.

If you've made it this far — not just through chapters, but through shifts in perception — thank you.

Not just for reading. But for listening.

Because this isn't the end of a book. It's the beginning of an alignment.

The Moon will not change after you read this. But maybe, just maybe, you will. And next time you look up...you won't see distance. You'll feel resonance.

– *Sandeep*

# Appendices

# Appendix A: Mainstream vs UED

## Comparison Table

*A Side-by-Side Breakdown of Paradigm Assumptions*

This comparison isn't meant to antagonize conventional science — it's meant to liberate inquiry. Where the mainstream sees mechanics, UED sees meaning. Where the classical model explains through motion, UED reinterprets through ripple alignment and field posture.

| Phenomenon / Concept | Mainstream Science | UED (Universal Energy Dynamics) |
|---|---|---|
| Motion | Bodies move through space due to force, momentum, or inertia. | Apparent motion is a ripple phase shift within a still field. Bodies don't "move" — field patterns shift. |
| Gravity | Mutual attraction between masses, decreasing with distance. | Field compression due to ripple imbalance. No "pull" — only tension returning to equilibrium. |
| Tides | Caused by gravitational pull from Moon (and Sun), with delay. | Result of ripple breathing and ATT (Absolute Threshold of Tension) loops. The Moon stabilizes field tension. |
| Lunar Motion | Moon orbits Earth and is tidally locked. | Moon is posture-locked, not orbiting. Its fixed face is a result of ripple synchronization, not spin + orbit. |

| Phenomenon / Concept | Mainstream Science | UED (Universal Energy Dynamics) |
|---|---|---|
| **Earth's Rotation** | Earth spins on axis every ~24 hours. | No objective spin — day-night cycles are caused by field phase loops. Spin is perceptual, not mechanical. |
| **Seasons** | Caused by Earth's axial tilt and orbit around Sun. | Ripple intensity shifts between hemispheres. Seasonal rhythms arise from field exposure, not tilt or revolution. |
| **Galactic Rotation** | Galaxies rotate; outer stars remain inexplicably fast (dark matter invoked). | No spin — galaxies pulse in ripple cycles. Spiral arms are standing wave zones, not material trails. |
| **Time** | A measurable, flowing dimension. | A rhythmic effect of ripple transitions. Time is not flowing — it's felt as coherence through ATT cycles. |
| **Perception** | Passive observation by an external observer. | Active participation in field resonance. Perception arises when observer aligns with ripple frequency. |
| **Moon's Role** | A gravitational satellite influencing tides and eclipses. | A field stabilizer — the Moon maintains ripple symmetry, enabling rhythm, breath, and consciousness. |
| **Black Holes** | Dense objects with infinite gravity from collapsed stars. | Ripple anchors with maximum compression. They hold field posture — not devour matter. |

| Phenomenon / Concept | Mainstream Science | UED (Universal Energy Dynamics) |
|---|---|---|
| Expansion of Universe | Redshift interpreted as galaxies moving apart — space expanding. | Redshift = phase delay, not motion. The universe isn't expanding — it is oscillating through ripple cycles. |
| Forces (EM, Gravity, etc.) | Separate interactions mediated by particles or spacetime curvature. | All forces are field consequences — not causes. Ripple misalignments cause observable "force effects." |

**Final Insight:**

Mainstream science measures the visible.

UED listens for the invisible rhythm that enables it.

The difference is subtle — but transformative. We don't discard Newton or Einstein. We complete them — by anchoring their models within the field that was always there.

Stillness is not a rejection of science. It is its recalibration.

# Appendix B: Testing the Ripple Model

## Observational Tools and Suggested Experiments

While Universal Energy Dynamics (UED) reframes much of mainstream physics, it is not merely philosophical. It opens the door to observable, replicable, and refinable methods of investigation — especially when guided by field awareness rather than mechanical assumptions.

Below is a curated set of tools, techniques, and experiments to explore ripple-based phenomena practically.

**I. Visual Observation: The Moon's Stillness**

**Tool**: Long-term time-lapse photography or astrophotography (with Earth-referenced frame) What to Observe:

- The unchanging face of the Moon
- Apparent lack of rotational drift over months or years
- Phase transitions as perceptual shifts, not orbital causes

**Suggested Question**: Can motionless positioning still explain rhythmic lunar phases without invoking rotation or orbit?

**II. Ripple-Driven Tides: Testing ATT Response**

**Tool**: Coastal tidal monitors, synchronized with lunar ATT models What to Measure:

- Tidal highs/lows vs lunar ATT cycle prediction, not just gravitational timing

- Delay patterns in tidal peaks during lunar resonance windows
- Landlocked body fluctuations (e.g., enclosed lakes, aquifers) during full/new moons

**Suggested Question**: Are tides consequence patterns of ripple breathing rather than gravitational pulls?

**III. Biological Correlation: Moon–Mind Synchrony**

**Tool**: EEG sensors, heart rate monitors, menstrual/fertility tracking apps What to Correlate:

- Human brainwave patterns and emotional states during different lunar phases
- Circadian irregularities during lunar extremes
- Reproductive cycle shifts across waning vs waxing phases

**Suggested Question**: Is human physiology entrained not to visual moonlight, but to invisible ripple field resonance?

**IV. Eclipses as Interference, Not Shadow**

**Tool**: Polarized light sensors, EM flux detectors, live eclipse recording (solar/lunar) What to Test:

- Sudden field shift patterns (e.g., voltage fluctuations, EM disturbances) during eclipse onset
- No measurable "shadow" on instruments — only phase cancellation or attenuation
- Behavioral anomalies in animals or humans during eclipse (field confusion effects)

**Suggested Question**: Can an eclipse be identified through field interference instead of geometric light blocking?

**V. ATT Mapping: Earth–Moon Resonance Zones**

**Tool**: Scalar field probes, sensitive magnetometers, consciousness coherence platforms (like Global Coherence Initiative sensors) What to Map:

- Lunar ATT resonance zones where emotional, biological, or geophysical effects intensify
- Field tension curves over time, in relation to lunar phase and observer alignment
- Points of enhanced intuition, synchronicity, or coherence

**Suggested Question**: Can we locate and time ripple harmonics the way we map gravity or weather?

**VI. Tuning Experiments: Local Ripple Field Sensitivity**

Simple DIY Tools:

- Tuning forks in sealed environments
- Fluids in still containers (observing harmonic agitation)
- Metal pendulums left suspended during full/new moons
- Sleep logs, dream intensity tracking, meditation depth journals

What to Track:

- Field agitation without physical stimulus
- ATT thresholds triggering subtle rhythmic behavior in matter and mind
- Synchronous emotional or physical shifts across individuals in unrelated locations

**Suggested Question**: Can a change in planetary field posture produce observable consequences without motion or applied force?

**Closing Thought: You Are the Instrument**

Mainstream experiments require devices. UED invites direct participation.

The Moon's message can't always be picked up with sensors — because you are the sensor.

To feel ripple alignment, to detect field shifts, to validate rhythm over force — you must learn to observe not just with your tools, but with your own coherence.

*The lab is the sky. The method is stillness. The subject is you.*

# Appendix C: Lunar Phases and ATT Chart

**Ripple Thresholds in the Rhythm of the Moon**

The phases of the Moon are not merely visual changes caused by reflected sunlight — they are field expressions of Earth–Moon ripple alignment. In UED (Universal Energy Dynamics), these phases are not passive reflections, but ripple posture states that correspond to different ATT (Absolute Threshold of Tension) values within the Earth–Moon system.

Each phase marks a specific field configuration where tension either builds, releases, stabilizes, or resets.

**Understanding ATT: Absolute Threshold of Tension**

In UED, ATT represents the point at which ripple field tension reaches a structural turning point — allowing:

- Energetic release (e.g., tides rising or falling),
- Biological entrainment (e.g., sleep, emotion, hormonal cycles),
- Perceptual clarity or distortion (e.g., during new/full moons or eclipses).

These thresholds are not determined by light or gravity — but by phase coherence and ripple resonance.

## Lunar Phase – ATT Chart Overview

| Lunar Phase | UED Ripple State | ATT Behavior | Field Consequence | Observed Effects |
|---|---|---|---|---|
| New Moon | Ripple compression node | High inward ATT – minimal outward release | Field resets; introversion; stillness spike | Low tides; dreamlessness; lowered emotional surface |
| Waxing Crescent | Gradual phase expansion | ATT slowly rising | Field stabilizes for outward expression | Emerging energy; productivity boost; cellular activation |
| First Quarter | Midpoint pressure balancing | ATT threshold breach | Action phase; decisive behavior; ripple output peaks | Tidal uplift begins; heightened motivation or conflict |
| Waxing Gibbous | Pre-maximal ripple build-up | High outward tension, nearing release | Preparation for full release | Body strain; anticipation; premonitions; high sensory activity |
| Full Moon | Ripple expansion max | Peak ATT – major tension release | Overflow of emotion, thought, and field resonance | High tides; insomnia; vivid dreams; heightened awareness |
| Waning Gibbous | Controlled dissipation phase | ATT starts falling | Reflection and integration | Clarity; digestion; cognitive grounding |
| Last Quarter | Tension neutral reset | Stable low ATT | Energy reorganizing toward inward stillness | Emotional neutrality; intuitive decisions |
| Waning Crescent | Return to still posture | Lowest ripple fluctuation | Deep rest and inward orientation | Best sleep; renewal; subconscious processing |

## ATT Cycle Interpretation Guidelines

- **High ATT (Full/New)**: Times of consequence emergence — events, decisions, shifts.
- **Low ATT (Waning)**: Phases of internal integration, healing, and subconscious tuning.
- **Threshold Points (Quarter Phases)**: Moments of internal

field conflict, great for action or redirection.

**Applications of the Chart**

1. **Health & Behavior**: Align sleep, fasting, or creative flow with ATT-sensitive windows.
2. **Decision Making**: Initiate during waxing phases, consolidate during waning ones.
3. **Meditation & Rest**: Deepest internal coherence occurs just before or after New Moon.
4. **Tidal Mapping**: ATT aligns more precisely with oceanic swell and recession than classical gravity predictions.

**Final Insight: The Moon Does Not Light the Night — It Times the Field**

In UED, lunar phase is not a clockface of light — it is a language of field posture. ATT is how the universe inhales, holds, and exhales — and the Moon shows us where we are in that breath.

# Appendix D: Glossary of UED Terms

## A New Vocabulary for a Still Universe

Universal Energy Dynamics (UED) introduces a paradigm shift in how we understand physics, perception, and existence. This glossary serves as a bridge between classical terminology and ripple-informed awareness, helping readers realign their language to match a universe built on structured stillness, not motion.

**Absolute Threshold of Tension (ATT)**

A specific limit within a ripple field beyond which a consequence manifests — similar to how water must reach a certain temperature to boil. In UED, ATT defines when a ripple becomes visible, a field collapses into form, or a tide, emotion, or idea surfaces. It governs timing, perception, and phase shifts.

**Ripple**

Not a wave, not a particle — a localized misalignment within structured vacuum energy (SVE). A ripple carries consequence without carrying mass, and becomes the base unit of action in the universe. Everything from light to thought to gravity is a ripple behavior.

**Ripple Loop**

A recurring field pattern where ripple consequences move through a closed cycle. What appears as orbit, rotation, or oscillation is actually a ripple loop — repeating due to posture-locked phase harmonics, not motion.

**Posture Lock**

A field state in which an object (like the Moon) holds a fixed configuration relative to ripple dynamics. It does not move but remains resonant with its field — creating the illusion of motion (e.g., rotation or orbit) from an observer's changing perspective.

**Structured Vacuum Energy (SVE)**

The still foundational fabric of the universe, out of which all ripple activity emerges. Unlike empty space, SVE has tension, structure, and symmetry — allowing for alignment, resonance, and coherence. It is the true "medium" behind all physical and perceptual reality.

**Field Compression**

What classical science interprets as gravity is redefined in UED as zones of higher ripple tension — not mutual attraction between masses, but a drive toward field equilibrium. Compression increases consequence expression (e.g., tides, dense cores, awareness spikes).

**Tension Geometry**

The configuration of field stress at any point in space. Rather than measuring force vectors, UED measures how ripple patterns build, cancel, or reinforce each other. Phenomena like eclipses, tides, and seasons arise from changes in tension geometry, not mechanical mechanics.

**Phase Coherence**

The state in which multiple ripple fields align in rhythm and posture. When coherence is high, clarity, stillness, and harmony emerge — in biology, behavior, climate, or perception. When coherence is low, misalignment causes instability, noise, or chaos.

**Consequence over Cause**

A core UED principle: the universe does not operate on forward-driving causality. Instead, consequences arise when structured tension thresholds are crossed. You don't need to push a system — you wait for alignment, and consequence flows from it.

**Resonance Window**

A moment in ripple dynamics where the observer and the field align precisely, allowing access to information, insight, or transformation. Lunar full moons, eclipses, solstices, and meditation moments are examples of resonance windows.

**Ripple Breathing**

The rise and fall of field tension in biological and planetary systems — experienced as tides, emotional surges, or circadian rhythm. It is the pulse of consequence in the absence of external motion. The Earth breathes through ripple cycles.

**Temporal Illusion**

Time is not absolute, nor linear. In UED, time is the effect of ripple phase transitions. What we call "passing time" is our body-field moving through shifting ATT landscapes — much like moving through seasons in a field that never left.

**Stillness**

Not the absence of motion, but the intelligence that precedes it. In UED, stillness is the most active presence — a structured field that holds everything in balance, allows rhythms to emerge, and reveals meaning without moving.

# Appendix E: Quotes from Newton Reconsidered

## The Moon That Undid Newton – In His Own Language

This appendix brings together imagined reflections and reinterpretations of Newton's ideas, as reframed through the lens of Universal Energy Dynamics (UED). These quotes are not direct citations from Newton's historical writings, but rather symbolic reinterpretations — showing how the very logic Newton helped shape now invites its own transformation.

Each quote below is paired with its UED reconsideration.

*1. "If I have seen further, it is by standing on the shoulders of giants."*

**UED Reframe:**

But if the view is distorted, even giants may look like dancers in motion — when all along, they stood still in perfect posture.

*2. "To every action, there is always an equal and opposite reaction."*

**UED Reframe:**

To every ripple, there is a resonance — not opposition, but alignment seeking homeostasis. Reality isn't reactive. It is rhythmically responsive.

*3. "Gravity explains the motions of the planets, but it cannot explain who sets the planets in motion."*

**UED Reframe:**

Perhaps nothing was ever set in motion. What you called gravity was structured stillness correcting imbalance — not force, but consequence.

4. *"The Moon is continually falling toward the Earth, but never hits it."*

**UED Reframe:**

The Moon is not falling — it is anchored. It holds posture so precisely that we mistake its resonance for freefall.

5. *"I feign no hypotheses."*

**UED Reframe:**

Then let us offer no forceful hypotheses either — only field clarity. Stillness doesn't need feigning. It simply needs perceiving.

6. *"Nature is pleased with simplicity. And nature is no dummy."*

**UED Reframe:**

And the simplest truth may be this: what doesn't move may govern all that does. Stillness is not the absence of complexity — it is its quiet master.

7. *"Truth is ever to be found in simplicity, and not in the multiplicity and confusion of things."*

**UED Reframe:**

Multiplicity is the illusion of movement. Simplicity is the posture from which rhythm unfolds. The Moon never confused us — we misread its silence.

8. *"The motions which the planets now have could not spring from any natural cause alone, but were impressed by an intelligent Agent."*

**UED Reframe:**

That Agent is not external. It is the field itself — intelligent, coherent, and still. What you saw as impressed motion was embedded resonance.

9. *"I can calculate the motions of heavenly bodies, but not the madness of men."*

**UED Reframe:**

Because motion was never the right variable. Men lose alignment, not logic. The Moon never moved — and perhaps neither should we, until we first listen.

*10. "What we know is a drop, what we do not know is an ocean."*

**UED Reframe:**

Then let us ask — who made the ripple in that drop? For even a still ocean holds worlds not by waves, but by the breathless tension between them.

*Final Note:*

Newton's brilliance laid the groundwork for asking how the universe moves.

**UED now asks:** "What if it never did?"

And in that question, the Moon returns — not to orbit, but to teach.

# Appendix F: Ripple-Based Astronomy Framework

## Toward a New Science of the Sky

Traditional astronomy rests on a foundation of mechanical motion, where objects spin, orbit, attract, and accelerate through space. But if Universal Energy Dynamics (UED) holds true — and the cosmos is structured stillness, not chaotic propulsion — then we need a new astronomical model.

This appendix outlines the foundational pillars of Ripple-Based Astronomy: an observational, theoretical, and philosophical reframing of celestial dynamics based not on force or motion, but on ripple alignment, field posture, and tension geometry.

**1. Posture Over Position**

In ripple-based astronomy, objects are not moving through space, but holding specific postural alignments within large-scale structured vacuum energy (SVE) fields. What we perceive as orbits, spins, or axial tilts are field relationships — not mass trajectories.

Example: The Moon is not orbiting the Earth — it is posture-locked in a ripple field. The Earth's shift makes the Moon appear to move.

**2. Ripple Loops Instead of Orbits**

Instead of elliptical paths carved by gravitational mechanics, celestial motion is reinterpreted as ripple loops — closed cycles of field tension that repeat based on ATT thresholds (Absolute Threshold of Tension).

Eclipses, moon phases, seasons, and even planetary alignments are consequences of looping ripple geometry, not mechanical positioning.

**3. Stillness at the Core of Movement**

Each celestial system has a stillness node — a body or region that anchors the ripple field. These anchors (like the Moon in Earth's case, or black holes in galaxies) appear static but hold the structure together.

The stillness is not passive; it is the source of rhythm, coherence, and resonance across systems.

**4. Field Compression as Gravity**

Where classical astronomy relies on mass-based gravitational pull, ripple astronomy describes gravity as field compression — the net effect of unresolved ripple tension. Celestial coherence comes not from attraction, but from harmonic pressure balance.

Dense objects form where ripple stress converges and stabilizes — not because mass "pulls" on other mass.

**5. Perception-Dependent Phenomena**

Many observable astronomical events — such as moon phases, retrograde motion, and redshift — are now explained as observer-ripple interactions, not as absolute trajectories.

What we see is influenced by our position in the ripple field, not by actual displacement of objects.

**6. Phase Synchrony over Gravitational Lock**

In ripple astronomy, bodies that seem tidally locked or synchronized (e.g., the Moon always showing one face) are doing so because of ripple phase-lock, not because of tidal gravitational braking.

The synchrony is built into the field structure — a natural harmonic, not a mechanically forced state.

**7. ATT Events Replace Keplerian Predictability**

Where classical astronomy uses Newtonian and Keplerian math to predict planetary positions and events, ripple-based astronomy focuses on ATT cycle mapping — identifying when and where ripple tension builds, peaks, and resolves.

Eclipse prediction, solstice timing, or bio-rhythm alignment can all be ATT-calibrated, not orbit-calculated.

**8. Cosmic Breath, Not Cosmic Bang**

Instead of a universe expanding from a single explosive moment, ripple astronomy envisions a cosmos that breathes — expanding and contracting in field cycles, not in one-way motion.

Redshift, galactic pulsing, and dark energy anomalies are all reframed as ripple rhythm delays, not velocity-based recession.

**9. Applications for Field-Based Navigation and Timing**

In future ripple-based systems:

- Space navigation could align with ripple flow instead of velocity thrust.
- Calendars could use lunar ATT cycles instead of solar days.
- Human health could be regulated by syncing with cosmic field harmonics, not artificial time structures.

**10. Rewriting the Celestial Map**

Constellations, star charts, and sky atlases must eventually evolve to reflect:

- Field density zones instead of static object positions.
- Ripple harmonics instead of linear distances.
- Stillness nodes that organize surrounding behavior through posture.

**Summary Chart: Classical Astronomy vs Ripple-Based Astronomy**

| Aspect | Classical View | Ripple-Based View (UED) |
|---|---|---|
| **Motion** | Objects move through space | Ripple loops create perceived motion |
| **Gravity** | Mass-based attraction | Field compression from ripple tension |
| **Time** | Linear, clock-based | Phase cycles through ATT thresholds |
| **Rotation & Orbit** | Spin and revolve due to angular momentum | Illusions caused by posture-lock and ripple shift |
| **Eclipses & Seasons** | Line-of-sight blockage / axial tilt | Phase alignment within field structures |
| **Black Holes** | Infinite mass pulling in matter | Maximum compression nodes anchoring structure |
| **Redshift** | Light stretched by recession | Delay in ripple phase resolution |

**Final Note:**

Ripple-Based Astronomy invites us not to map what moves, but to feel what holds. The sky is not a stage of orbiting actors. It is a stillness symphony — and the Moon was the first note.

# Further Reading and References

## To those who wish to explore beyond motion — into meaning.

This book reinterprets celestial and physical phenomena through the lens of Universal Energy Dynamics (UED) — a framework proposing that stillness, not motion, underlies the cosmos. While much of this work stands independently, it also builds upon centuries of human curiosity, from classical science to metaphysics, from empirical observation to ancient wisdom.

Below is a curated list of foundational texts, complementary works, and sources of inspiration — both mainstream and unconventional — that readers may explore to understand the historical context, the contrast, and the evolution of thought leading to ripple-based cosmology.

**I. Classical Foundations (for context and contrast)**


- Isaac Newton – *Philosophiæ Naturalis Principia Mathematica*
- Albert Einstein – *Relativity: The Special and the General Theory*
- Galileo Galilei – *Dialogue Concerning the Two Chief World Systems*
- Johannes Kepler – *Harmonices Mundi (The Harmony of the World)*
- James Clerk Maxwell – *A Treatise on Electricity and

*Magnetism*

These works laid the groundwork for mechanical cosmology, force-based physics, and space-time dynamics. Understanding them helps illuminate the paradigm UED seeks to transcend.

**II. Disruptive Thinkers and Unresolved Questions**

- Nikola Tesla – Collected Lectures and Patents on Resonance and Energy Fields
- Halton Arp – *Seeing Red: Redshifts, Cosmology and Academic Science*
- Rupert Sheldrake – *Science Set Free*
- David Bohm – *Wholeness and the Implicate Order*
- Thomas Kuhn – *The Structure of Scientific Revolutions*

These voices raised key challenges to orthodoxy — asking questions about cosmological assumptions, experimental dogma, and the nature of scientific revolutions.

**III. Ancient Echoes and Field Wisdom**

- Rigveda, Upanishads, and Surya Siddhanta – Ancient Indian texts that implied a non-mechanical cosmos, structured through rhythm, cycles, and balance.
- Lao Tzu – *Tao Te Ching*: Insights on stillness, flow, and harmony beyond logic.
- Mayans & Babylonians – Their lunar-solar calendars aligned with field patterns more accurately than many modern systems.

These are not scientific textbooks but field-sensitive perspectives that point to a deeper awareness of celestial resonance.

**IV. Related Works by the Author (for deeper continuity)**

- Energy in Action: The Soul Demystified
- The Earth Doesn't Move
- Vishwarup Darshan: The Complete Vision of Reality
- Manifestation: From Nothing to Everything
- The Quantum Courtroom
- The Still Universe Model
- The War of Thoughts series
- Cracks in the Known
- The Human Game
- The Divine Dilemma
- Chavan's Atomic Model
- Chavan's Theory of Everything
- Naked Truth: Everything and Else
- Chavan vs Einstein

These titles offer progressive development of UED, exploring its implications in physics, consciousness, cosmology, and human behavior.

**V. Empirical Sources and Observation Data**

- NASA Lunar Reconnaissance Orbiter (LRO)
- US Naval Observatory Lunar Phase Charts
- International Astronomical Union Archives
- Lunar Laser Ranging Experiment (Apollo Program)
- NOAA Tidal Data Sets & Satellite Oceanography Reports

While interpreted differently in this book, these datasets and instruments remain valuable for tracking observed phenomena such as:

- Lunar phases,
- Tidal shifts,
- Earth-Moon distance variation,

- Eclipse cycles.

## VI. Ongoing Independent Research Projects (Recommended for Field Enthusiasts)

- Electromagnetic Universe Proposals (EU Model)
- Consciousness and Field Theory Research (e.g., IONS, HeartMath Institute)
- Alternative Cosmology Groups & Forums
- Astrobiology & Lunar Rhythmicity Studies
- Biofield Mapping and Circadian Resonance Experiments

These initiatives, while not always aligned with UED, reflect the spirit of inquiry needed to move science beyond the gravitational box.

**Final Suggestion:**

Read not to confirm — but to question. Observe not to believe — but to feel.

And always remember:

*The Moon didn't come to orbit us. It came to align us.*

# Acknowledgments

This book was not born in a laboratory, nor under a telescope. It was born in stillness — the kind that arises when you stop seeking answers and begin listening to the questions the universe has already asked you.

I am deeply grateful to the silent teachers of my life: the night sky, the untiring Moon, and the rhythm of days and tides that never asked to be noticed, yet never failed to guide.

To my students — you are the mirrors that reflect back my curiosity, and the questions you ask often carry more weight than any answer I could give.

To my family, who may not always understand the path I walk, but still walk beside me — thank you for your patience, your quiet support, and your belief that even the strangest truths deserve their day under the Sun... or the Moon.

To my critics and skeptics — you sharpened this work far more than praise ever could. The weight of disbelief was the tension needed to build coherence.

To the thinkers who came before — Newton, Einstein, Tesla, Bohm, and all those unnamed by history — your models were not wrong, just unfinished. I honor your brilliance, even as I dare to rearrange it.

To the Moon itself — this book is not about you. It is because of you.

Finally, to the reader who made it to this page: thank you. Whether you arrived here through science, spirituality, skepticism, or silent wonder — your attention is the ripple that completes this resonance.

We are not alone in the cosmos. We are in tune — when we choose to be.

— *Sandeep Chavan*

# About the Author

Sandeep Chavan is an independent researcher, educator, and author of over 100 books spanning physics, philosophy, consciousness, and cosmology. He is the mind behind the Universal Energy Dynamics (UED) framework — a paradigm that challenges conventional notions of force, gravity, and motion by placing stillness, ripple fields, and structural equilibrium at the heart of all phenomena.

Founder of Gyrus Education (OPC) Pvt. Ltd., Sandeep has taught science for over a decade, inspiring students to move beyond memorization and into meaningful questioning. His work is not confined to classrooms or equations — it is a philosophical journey woven through observation, logic, and bold reinterpretation.

He writes not to impress the mainstream, but to awaken the overlooked — and his books have quietly reached thinkers, rebels, and seekers across the world.

*The Moon* is not just one of his most daring works — it is also among his most personal: a tribute to stillness, to forgotten truths, and to the unseen architecture holding life together.

To explore more of his works or get in touch, visit:

https://www.smashwords.com/profile/view/SANDEEPCHAVAN

# Don't miss out!

Visit the website below and you can sign up to receive emails whenever Sandeep Chavan publishes a new book. There's no charge and no obligation.

https://books2read.com/r/B-A-EPGPC-WYFSG

BOOKS 2 READ

Connecting independent readers to independent writers.

Did you love *The Moon: Co-Author of Life's Script*? Then you should read *The Earth Doesn't Move*[1] by Sandeep Chavan!

[2]

**What if the Earth never moved—because it never had to?**

We've all been told: the Earth spins, revolves, races through space at unimaginable speeds. Newton calculated it. Einstein bent time around it. Satellites "prove" it. And yet... what if all of it was a story stitched from perception, not truth?

In this audacious book, **Sandeep Chavan**, creator of the **Universal Energy Dynamics (UED)** framework, dismantles the deepest assumptions of modern physics—starting with the motion of Earth itself. With sharp logic, field-based reasoning, and ripple-centered insight, he invites readers to reconsider everything from gravity to inertia, time to space, light to consciousness.

---

1. https://books2read.com/u/mgd5j7
2. https://books2read.com/u/mgd5j7

"Stillness is not the absence of motion. It is the condition from which all motion becomes visible."

Through 28 groundbreaking chapters, the book unfolds a radical thesis: **Earth is still**, and what we perceive as spin, orbit, force, and expansion are ripple consequences of field misalignment—not actual movements through space. With detailed reinterpretations of Newton's laws, Einstein's relativity, and even quantum phenomena, the author shows how mainstream science has mistaken **consequence for cause**, layering explanations upon a false foundation.

This is not conspiracy, denial, or mysticism. It is rigorous reasoning outside the conventional lens. It is physics, philosophy, and perception reconciled through the clarity of **structured vacuum energy** and **ripple alignment thresholds**—cornerstones of the UED view.

Whether you are a student, a scientist, a philosopher, or just someone who looks at the stars and wonders why they never move—this book will awaken something deeper. It won't just tell you what to believe. It will make you **see** again.

You will learn:

Why day and night don't require Earth's spinHow inertia is ripple-lock in still fields—not resistance to motionWhy gravity is pressure, not pullHow satellites orbit without Earth revolvingWhy time dilates not due to speed, but ripple threshold delaysHow consciousness is not separate—but the very ripple that perceives

**Question what moves. Reclaim what is still.**

**Dare to look again—not at the sky, but at the belief behind your eyes.**This is not the end of science. It's the return to real observation.

Read more at https://www.sandeepjchavan.com/.

# Also by Sandeep Chavan

**Chavan's UED research**

CMGED Research Report
Chavan's Cosmic Continuum ($C^3$)
Frequency
Ripples of Reality
Reality Realigned
The Great Realignment
I – The Cosmic Leverage
Chavan's Cosmic Theory of Relativity (CTR)
Chavan's Theory of Light
Feeling Like a Physicist
TIME – The New Definition
$E \neq mc^2$: The Collapse of Classical Assumptions
The Chavan Codex
Chavan's Atomic Model
Chavan's Theory of Everything

**Echoes of War: The Russia-Ukraine Conflict and Its Global Ripples**

Shattered Lives
The War Economy
Borders Redrawn

Arms and Allies
Collateral Damage
Justice in the Ashes

**Somewhere in the Darkness of the Cosmos**
The Blueprint of the Universe
The Cosmic Breath
From the Edge of the Universe

**The Eternal Sage Chronicles**
The Sage and the Sovereign
The Family Pillar
The People's Flame
The Warrior's Soul
The Bridging Mentor
The Guiding Light

**The Exodus**
The Last Cycle: 2075

**The Fundamental Theory of Life & Existence**
Energy in Action: The Soul Demystified
The Active Mind
Energy in Connection
Death: The Disconnection
Subconsciousness: The Chosen Disconnection

Desires: The Engine of Life

**The Restoration of Human Knowledge**

Restoring Bhagavad Gita

**The War of Thoughts**

Cracks in the Known

The Human Game

The Divine Dilemma

The Original Intent

The Answers

**Ultra Modern Life Philosophy**

The Duality of Laughter and Tears

Take It Easy

Compare to Accept

**Standalone**

It's Not AI, It's AHI - Amplified Human Intelligence

The Decision Paradox

The Four Sapiens

Malicious Script of Indian Polity

The IIT Legacy & Global Impact

The IIT Legacy & Global Impact

You are just a Version of Your Original

Win the Game You Didn't Choose

Always Keep Your Bags Packed
Reveal to Shield - Learn the Game of Facades
Engage Beyond Elections
Culture, Identity & Change: The Evolution of Indian Society
The Triangular Dynamics
Never Trust Without Doubt
If You Can't Change the Country, Leave the Country
Live Fearless with Fear
India: The Investment Magnet
Realignment
Powering a Billion Dreams
Universal Skill
Extension: From Brush to Algorithm
Human Side of Einstein
The Future Knows No Lines
Let's Bring the Generations to the Table
Unified Energy Dynamics
A Beat That Never Dies
Strategic Horizons
Right From Scratch
The Forgotten Innovators
Physics Fixing
The Savior Illusion
The Selfish Side
The Age of the Weakest Sapiens
Neural Wars
The Absorption: There Is No Death
Strange Therapist
Bubbles: Become a Drop
Beyond the Blue
Half-Scholar
The Law of Truth
Failure

Philosophy 2.0: A Structural Manifesto
Chaos: A Gentle Guide to the Unresolvable
Self-Help: New Syllabus
Rippling
Game: How to Play the Share Market
Out of Syllabus
Saffronized: The Code Behind the Crown
The Anger
Moksha: The Liberation
Chavan vs Einstein: The Universe Was Never Curved
Rocket Science: The Ripple Gambling
The Cosmos: The Still Universe
The Taste of Silence
Chavan's Post-Materialist Human Model
10-Second Rule
Chavan vs Hawking
It was Never a Wave Signal
Quantum Courtroom: The Trial of a Century
Chavan–Heisenberg Ripple Uncertainty Principle
The Great Disruption
Dark ≠ Hidden
Breath
Calculus of Life
The Earth Doesn't Move
Black Hole Don't Exist
Chavan vs Kant
Naked Truth: Everything and Else
The Moon: Co-Author of Life's Script
World War-3

Watch for more at https://www.sandeepjchavan.com/.